Signs of Danger

Edited by

Sandra Buckley

Michael Hardt

Brian Massumi

THEORY OUT OF BOUNDS

Signs of Danger

Waste, Trauma, and Nuclear Threat

Peter C. van Wyck

Theory out of Bounds *Volume 26*

University of Minnesota Press

Minneapolis • London

For Ying Li

Figures 1, 6–8, 10–14, and 17 are from the U.S. Department of Energy report
Expert Judgment of Markers to Deter Inadvertent Human Intrusion into the Waste Isolation Pilot Plant,
by Kathleen M. Trauth, Stephen C. Hora, and Robert V. Guzowski
(Albuquerque: Sandia National Laboratories, 1993).

Published by the University of Minnesota Press
111 Third Avenue South, Suite 290
Minneapolis, MN 55401-2520
http://www.upress.umn.edu

LIBRARY OF CONGRESS CATALOGING-IN-PUBLICATION DATA

Van Wyck, Peter C.
Signs of danger : waste, trauma, and nuclear threat / Peter C. van Wyck.
p.   cm. — (Theory out of bounds ; v. 26)
Includes bibliographical references and index.
ISBN 0-8166-3762-8 (hc : alk. paper) — ISBN 0-8166-3763-6 (pb : alk. paper)
1. Radioactive wastes—Environmental aspects.   2. Radioactive waste
disposal in the ground.   3. Waste Isolation Pilot Plant (N.M.).
4. Threat (Psychology).   5. Psychic trauma.   6. Signs and symbols.
I. Title.   II. Series.
TD898.V36 2004
304.2'8—dc22
2004017298

Printed in the United States of America on acid-free paper

The University of Minnesota is an equal-opportunity educator and employer.

12 11 10 09 08 07 06 05   10 9 8 7 6 5 4 3 2 1

Contents

Preface

HAVING NOW reached the far shore of the twentieth century and portaged into the grainy noir of the twenty-first, nature itself is no longer where or what we thought it was. And perhaps it is not there at all. If the question of nature was an embattled region in the late twentieth century, it has by now passed into a realm of pure undecidability. And yet, as ruins and histories and bodies and leakage, each one material and narrative and discursive and memorial, the burden of the twentieth century persists and issues an imperative for an understanding that can make sense of the breakdown of things. And breakdowns there are, many of them.

Questions like Where is nature? or What is nature? have long lost any residual probity or innocence they may have had (or pretended to). I do not wish to lament this passage or to call for a return to a prior state or to resolve the question by providing an alternative geography of a nature in ruins—a lament by other means. But I do wish to judge the implications of living with a nature that is no longer what or where we once thought it to be. The point here is not that nature has become cultural; one might as well argue that culture has become feral. And in either case this would still depend on one term as the explanatory frame for the other. Rather, the point is that nature has become political. Whatever nature was, is supposed to be, could be, and so on, is not, for this reason, beside the point. Such questions continue to be a preoccupation of techno-scientific, social, and political endeavors. Yet if nature has slipped into the register of the political, it seems imprudent at best (and quite probably perilous) to also continue as though it can assist in the formulation of a radical politics, as though nature is a guide in contesting social practices and in the diverse endeavors through which constituencies seek various modes of protection, acknowledgment, compensation, and remediation. If nature is also a part of a political field, a contested and virtual object, it cannot also stand on the outside as an ontological limit. It derives its normative character not from the ontic, from the world, but from the sphere of human action. It does not whisper truth to us. It is not a reminder of what is possible, good, and desirable, and what is not. It cannot be these things. It cannot be the expressive force of a cultural counterimage nor can it be the given ground of human affairs.

The claim so central to ecological thought today is that everything is connected. Everything. As a claim, it is so deeply embedded in popular and scientific imagination that its curious implications are seldom explored. Indeed it seems so self-evidently true that there is little more to be said. It is just unquestioningly the case that everything is connected. This is seen in multiple forms and political settings. For example, it is in the notion of an organicist Gaia, and it is in popular ecological maxims such as "there is no such thing as a free lunch." It is very much part of the theoretical, communication-based work of Tony Wilden but also in more recent work, such as Verena Andermatt Conley's book *Ecopolitics*, that seeks to discover in ecology an unacknowledged foundation of poststructuralism (the repressed of the events of May).[1]

In the *oikos* everything is connected, as we are to it. Without this fundamental commitment to interconnectedness, ecological and environmental thought would be evacuated of most (if not all) of their radical potentials; they would become *conceptually* powerless.[2] The very notion of an ecosystem presupposes a dynamic interaction of correlates as an organizing feature. From the point of view of ecological theory proper (Odum, Margalef, et al.), interconnectedness is indeed less of a transcendental principle than it is something that is just empirically given. In other words, belief in it is always supported by evidence that was there in advance. Just read the signs.

The way that environmentalism and ecological thought have become discursively organized requires this interconnectedness of everything. This foundation provides the very possibility of contesting practices that are geographically and temporally remote or that otherwise appear to be causally unrelated. Interconnectedness allows us to build up a picture of interactions and allows this picture to be superimposed on technical and industrial practices that are otherwise construed as unrelated. As a concept it facilitates a kind of discourse that allows one to say that what happens *over there* makes a difference *over here*. And it has frequently done so with potent results and has been doing so more or less since the publication of Rachel Carson's book *Silent Spring*.[3] Her work marks a moment in time when the very equipment of causality changes, when causality itself becomes ecological. For example, it offered a fantastically rich model of causality whereby phosphates in household laundry detergents became intimately linked to such apparently remote systems as fish production and employment in the trucking industry. It drew lines between production and consumption that were radically new, and it did not take long for its normative possibilities to become realized. Equally, it made of ecology a formation approximate to a legal framework through which "accidents" such as Love Canal could be contested on behalf of dead, living, and, significantly, future victims. Indicator organisms, the introduction of exotic species, fisheries decline, global warming—such things are palpable evidence of how the epistemology of interconnectedness has fostered forms of knowledge that have and continue to challenge various technoscientific incursions. Indeed, from this point of view specifically, this text you now are reading can be considered a contribution to ecological theory.

There are, from a practical point of view (political, pragmatic, ethical), good reasons for advancing claims based on interconnectedness. What happens over there often does make a difference over here (although usually the opposite).

Yet the problem is this: contemporary ecological threats can come to make ecological thought itself look like a particularly advanced form of cultural paranoia. I mean this in the sense that once we say that everything is connected in this fashion, we mean that everything is, if not already, then at least potentially integrated into a framework of understanding. And it isn't. To make everything connected is to see the fissures and cracks rendered by ecological threats—whether the threats posed by wastes or the threats retroactively discovered through accidents— as a kind of recompense for a failure to have properly understood the connections. *The real punishing the epistemic for its sins of omission.* When we look at threats—for example, mad cow disease, global warming, HIV—and then when we look at events that follow from them, we can often see no correspondence. Or, and it comes to the same thing, we don't know what is supposed to correspond with what. It is as though we are given the world of threat as a kind of syntagmatic organization, an articulation in a strange grammar. We respond by attempting to understand the paradigmatic elements, those terms that are unified in their absence. All we can do is suffer under the crushing weight of events and cast looks of suspicion on our inadequate models.

As one brings a work to its conclusion, it seems inevitable to wonder if it is still important, if it still feels relevant. Yet as I look around today, I see not much more than a tremendous and troubling affirmation.

This work began as an attempt to think about a monument to nuclear waste. An insane proposal from the end of an insane century. This seemed more than enough. But as the work proceeded, it spread out considerably. The problem of nuclear commemoration in the desert had turned out to be far larger than I had thought. I was drawn into a much broader consideration of contemporary threat itself (nuclear, ecological). And this is a wide domain, obviously.

Thinking through the strange optic of the nuclear became a way to think about some dimensions of contemporary ecological threat in general. The twentieth century was staggeringly productive in this regard. And the current century looks remarkably similar in many respects. And in any case, these threats are very much with us still, and, it seems to me, ought to be matters of pressing concern for us. A rising ocean. A falling building. A toxified river. A disappeared species. A nuclear landscape. All of these representing completely different matters and forces gathered together uniquely. And each marked by features that render them powerfully and radically new.

To assess the implications of all of this remains a project of immense importance. This present work is only a beginning. While the twentieth century brought with it such a singularity of events, of catastrophe, it also revealed itself incapable of accounting for its events in a theoretically convincing fashion. Indeed, we have come to understand the previous century as a time that authored the *unspeakable*. The event that exceeded our capacity to understand it or to speak it. Even though the silence is sometimes political, sometimes shame, sometimes indifference, the traumatic event—at least from the academic domains that I tend to inhabit—is presumed to be *the* model.

My thinking about all of this began with the sense that we need to be taking seriously the perilous deficit in contemporary ecological thought and practice. An ecology tooled for war and spills. In other words, it became clear to me that the reason we end up with a spectacular nuclear dump in the desert is not because it will do the job required of it or because it's just a good idea, but precisely because no one in charge could think of any other way to think about the problem. Inertia, in other words.

There is as much war and genocide and disaster (natural and otherwise) taking place now as at almost any other point since World War II. Some would say more; I'm not sure it matters. How are we to take measure of this? What are the means by which one might judge? What grounds might judgment itself call on to secure it *as* a judgment? Risk? Freedom from terror?

And the threats posed by warfare that today grips a large part of the globe, how are we to understand these? The Gulf War, Part II, although by no means over, is attempting to reinvent itself as an imaginary humanitarian adventure in "democracy building." Here, "weapons of mass destruction" have come to play a role of grand, invisible MacGuffin, a retroactive pretext for a war that really *did* take place.

What purchase might discourses such as that of risk have on any of this? It's not just that there are interests and that because these interests are different things get interpreted differently. This is much too easy, and none too helpful, and leads us at best to an analysis of ideology. I think that the point is different. One must come at it from the other end, so to speak.

Félix Guattari summed this up well, when at the end of *Chaosmosis* he wrote:

> The contemporary world—tied up in its ecological, demographic and urban impasses—is incapable of absorbing, in a way that is compatible with the interests of humanity, the extraordinary techno-scientific mutations which shake it. It is locked in a vertiginous race towards ruin or radical renewal. All the bearings—economic, social, political, moral, traditional—break down one after the other. It has become imperative to recast the axes of values, the fundamental finalities of human relations and productive activity. An ecology of the virtual is thus just as pressing as ecologies of the visible world.[4]

Of course the difference here turns on more than the question of visibility, and for Guattari there isn't really an opposition between the two ecologies, but what seemed important to me was the sense that the virtual was a way to think about ecological threats. Such an ecology is in a sense an ecology of the strange. At least it is from the point of view of an episteme of full understanding, an episteme of interconnectedness and the possible. Yet from another perspective, an ecology of the virtual would be nothing of the sort. After all, the strange is something actual that seems unreal or out of place. The strange is like the Freudian uncanny *(unheimlich)*, the uncanny, something familiar but foreign ("the *unheimlich* is what was once *heimisch*, home-like, familiar"[5]). The strange requires the home, an *oikos*, as a reference and index to its very strangeness. That is, its relationship, its interconnection with the *oikos*, remains both necessary and obscure. An ecology that could attend to the threats of the nuclear would have to make peace with the incomprehensible (and deadly) creativity of the virtual (and indeed creativity *is* the word here). Yet at the same time, we can see that there are good reasons to be wary of the move to call for the articulation of yet another new ecology, something I have done myself in *Primitives in the Wilderness*.[6]

Ecological threats and nuclear wastes in particular are perhaps not a call to understanding at all—at least not a boundless understanding of connections or presence. Perhaps the kind of understanding that is called for is that we can come to know such threats only indirectly, only perhaps through our responses to them (here we would ask questions about social meanings and values) and, above all, only through a radical reappraisal of the discourses of ecology and risk that have come to operate as the clearinghouse for questions of social well-being. Visible ecologies cannot tell us much about 10,000 years in the future. But then again, neither can Guattari's ecology of the virtual. Perhaps it is that the whole question of nuclear waste is more about augury than ecologies of the visible and the virtual. Or perhaps it is just about survival. In any case, I have tried to make my text less anchored in an insistence on interconnection and more alive to a world in which the web of filiations of the *oikos* is the only model for making sense of threat.

Walter Benjamin alerted us—although at a different point in time—to an image of the *Angelus Novus*, Paul Klee's (contemplative) angel of history. Benjamin offers us this angel as a redemptive figure. This angel of history would like to restore the past to a wholeness, to make reparations, to awaken the dead. But it can't. The angel might want to stop and help us, but the storm has caught its wings. This angel, whose face is ever turned toward the past, can see only one single catastrophe hurling wreckage at its feet. So it goes.[7] Yet one might wonder if it is possible that the angel is pointing the wrong way.

The idea for this work dates back some number of years to a very late-night encounter with a singularly fascinating article in *Harper's Magazine*. The piece was by Alan Burdick and is entitled "The Last Cold-War Monument."[8] The problem it described has provided me with a rich and productive tableau upon which to pursue my various obsessions—academic and otherwise. Although he and I have never met, nor, as far as I know, is he aware of this work, it seems that in a very impersonal personal way he must occupy a special place in the list of those to be "acknowledged."

As with any piece of work there are always others—some living, some dead, some present, others absent—who have made a contribution (in ways, of course, they may not recognize at all). With apology in advance for omission, my list is thus: Jonathan Bordo, Brian Massumi, Sandra Buckley, Lawrence Hazelrigg, Will

Straw, Charles Levin, Marike Finlay-de Monchy, Charles Acland, Maurice Charland, Vivian Patraka, Donald Callen, Kathleen M. Trauth, the late Thomas Sebeok, Stacey Johnson, Dann Downes, Tracy Marks and Anthea Browne of the Puntledge Kinesthetics Research Institute, Douglas Stilton Cohen, Michael Keene, Lester and Luigi of the Cedar Cottage Institute of Performance Textiles, Glenn Macdonald, Steven Butler, all the helpful staff at the Gordon Thompson Library, the late Duff Gordon, the Rathaus folks, the also late Diva, the now familiar yet still uncanny voice of "Fred," RA extraordinaire Joel McKim, all my supportive colleagues at the Department of Communication Studies at Concordia University, Balto the Wonder Dog, and most of all my partner Angela Aldinucci.

This work has twice benefited from the financial support of the Social Sciences and Humanities Research Council of Canada.

Norway Point, April 2004

Introduction

The sign does not wait in silence for the coming of a man capable of recognizing it: it can be constituted only by an act of knowing.

Michel Foucault, *The Order of Things*

The Surprising Fact

March 26, 1999. CARLSBAD, N.M.—Energy Secretary Bill Richardson today announced that the first shipment of defense-generated transuranic radio-active waste arrived safely at the U.S. Department of Energy's (DOE) Waste Isolation Pilot Plant (WIPP). Hundreds of people were on hand to watch this important milestone in the Energy Department's work to permanently dispose of defense-generated transuranic waste left from the research and production of nuclear weapons.... "This is truly a historic moment—for the Department of Energy and the nation," said Secretary Richardson. "This shipment to WIPP represents the beginning of fulfilling the long-overdue promise to all Americans to safely clean up the nation's Cold War legacy of nuclear waste and protect the generations to come."[1]

THE PLACE is Carlsbad, New Mexico. A very large hole has been excavated deep within the hard indifference of the desert's sedimentary salt. It is the world's first permanent, underground disposal facility for nuclear waste; it is stunningly expensive and equally controversial. In 1999, the Department of Energy, under the auspices of the government of the United States, approved the transport of transuranic (and thus very persistent) nuclear waste into this hole.

Sometime around the year 2035, the hole will be filled to capacity and sealed shut. And then the most extraordinary series of events will begin to take place. A series of events that captivate the imagination. By decree of the government a very large monument—in keeping with the magnitude of the burial beneath—must be constructed to mark the site. It will be perhaps the largest public works project in modern history. But this marker, this gravestone monument must serve both more and less than a commemorative purpose. Indeed this monument must seek to *not* commemorate. For what lies beneath must never be celebrated, yet in some fashion must always be remembered. The expenditure of the monument must be equal in magnitude to the waste contained beneath it. It thus cannot be a typical monument. It cannot be allowed to content itself as a monument to the present; it is not something that we wish to remember, nor is it something for which "we" wish to be remembered. It must, and again by decree, convey a very specific message to the future—and the message it must convey is Go Away!

It must be a calling to remembrance that celebrates nothing. *Look!, here lies nothing.* It must convince the future of its utmost significance and of its terrible danger. It must model the double movement of the burial. That is, on one hand the waste is made to disappear from sight, and thus is given over to safety. And on the other, the waste that lies below is to be given back to danger through the work of the monument. The material, hidden from sight, must be given back to danger by the sign. The register of the visual and the material idiom of the monument must convey the visceral.

This monument to signification must perform the threat that lies beneath. And it must do so for a legislated period of 10,000 years—say, around the year 12035. An ordinal number that should not be allowed to conceal its deeply cardinal implications. Three hundred generations. A Y12K problem.

It is a singular meeting of the material and the semiotic. And it is an enormous wager that hinges on making the waste safe—through burial—then making it dangerous again—through signification. And in it must persist the groundless hope that the semiotic decomposition of the sign will take place at a slower rate than the nuclear decomposition of the waste. The sign must outlive the waste; a question of half-lives (waste vs. meaning).

Of course this all may seem quite fantastic and ill-conceived. That may be so, but there is more to the story. As I began thinking about this project some years ago, it seemed to me to involve a number of problems related to questions of ecological thought. But it also seemed to point to something else—something that engages with profound questions concerning conventional notions of culture and nature, concepts of risk and objectivity, cultural conceptions of time and history, and the very idea of communicativity. The threats posed by the materials slated for burial are of a very particular sort. At once real and fictive—such threats bear witness to the erosion of boundaries by which a culture (which is to say European culture) has come to recognize its own precinct as distinct from that of nature. They are threats of a properly, or paradigmatically, ecological sort.

Ecological threats are awkward and dangerous and lively, and very difficult to picture. They cannot be adequately contained within an arithmetic of risk and probability. As ecological threats they threaten the very basis of what supports organic life, and they threaten too the very symbolic universe within which threat itself has meaning. The organic *and* the symbolic—ontological threat. Such threats are, therefore, irreducibly ethical as well.

The ethical concern of this work goes well beyond the "moral" obligation on the part of peoples currently or historically engaged in nuclear technologies to confront the threats posed by such nuclear practices. The threats posed by these activities (and their various residues) simply do not conform to traditional notions of responsibility and reparation, or location and jurisdiction, or for that matter, cause and effect. Ecological threats do not neatly cut along lines of class or location, victim and perpetrator, and do not adhere to assumptions about sovereignty or geopolitical regions. Unlike previous industrial threats that could be located and circumscribed, understood and written off, ecological threats are chimerical.

The Measuring Rod

Imagine the perplexity of a man outside time and space who has lost his watch, and his measuring rod, and his tuning fork. I believe, Sir, that it is indeed this state which constitutes death.

Alfred Jarry, *Exploits and Opinions of Dr. Faustroll, Pataphysician*

Figure 1 is an image of a design for the desert monument. Do you understand?

It's all happening at the edges. At the edge of the social, at the edge of the imagination. At the edge of memory, and the probable. No bodies, nobody.

Figure 1. Spike Field marker design. Concept by Michael Brill and art by Safdar Abidi.

There is no one there. Except, that is, in the drawings of the design ideas. There are quite a number of them. Tiny humans at the cemetery for waste.

True, it is a convention of sorts to do this—in certain kinds of documentary photography, that is. Sometimes it's a scale of some kind, a ruler propped up along the cleanly excavated face of an archaeological dig—there to show the size of the object pictured. And sometimes it's just an appendage, or miscellaneous article of use. But in any case, the intention is to characterize anthropomorphically the magnitude of an object, and in doing so to inject into it a comparative and documentary veracity. This is what allows this kind of image to show (Wittgenstein)—a measuring rod by other means.

But here, in these images, here we see little humans, as if to ground not only the magnitude of the monuments but to give them a place in the world. But here, what is shown is nothing as simple as the size of a pottery shard, or a length of a femur, but the very fact that these images are to be understood as part of a world, our world.

In addition to the measuring rod, in other words, the designs seek also a watch and a tuning fork. Without them, they would seem as fantasy drawings culled from an archive of Roger Dean album artwork. So they are there, these humans, there where they are not supposed to be.

To take the human away would be to present the unthinkable possibility that no one might be around to see it. It is—I think—"psychologically" necessary to show how a human might relate to the monstrous fact of these installations. The images need the uncanny to make them intelligible. The human must be there as the index of the inhuman, as the zero degree of the strange. The home, or at least an *oikos*, is required in order that the strange may find its place elsewhere. The strange of the monuments requires a home to convey their very strangeness.

And they sure are big; of this the little humans attest. So fragile, and so small somehow. So much so that the little humans are perhaps the only choice for a scale. What else would work? There is monumental duration on the one side, finite mortality on the other. The monument stands for a geological permanence. And the human for its dutiful respondent.

How can one possibly guarantee the other? In one sense they must. For it is precisely the task of the monuments to keep humans away, and the task of the humans to keep away from the site.

You don't have to look far to discover a problem; you just have to look. One *sees* the problem. What *are* they doing there? The problem operates in an immediate and visual register. How else? These monuments are haunted by their own failure, even in advance of their construction. As *ideas*, they carry the humans into the region from which they are to be excluded. Humans are parasitically present within the very gesture that would seek to guarantee their absence. As with other forms of contemporary wilderness, the human is always present, one way or (usually) another.

But even in this, there is a strange visceral (or strangely visceral) sense of having one's body there. For me it is the little human that stands before the Menacing Earthworks. I would like to be there. There to witness the place in the desert where this will take place. And it will *take* the place. Taking it out of a series of "present" moments into a distinct series of future. Not history as the site of the now (Benjamin), but the present as the site of the "not yet." This is the particular messianism of threat. Not yet.

This is not to say that the threat is somehow distant, pointing to a future only; as if it wasn't both immediate and present. It is. The point is that the threat—*as threat*—has the status of a paradoxical event. On one hand, it is something that is in advance of the accident, something in advance of *that which befalls*. But on the other, to be under threat is for something to already have taken place. To be under the threat of nuclear contamination is for *many* things to have already taken place. This is key, and it is in part what distinguishes contemporary technoscientific endeavors from their industrial precursors.[2]

The rise of contemporary technological practices brings correspondingly more complex accident scenarios. But the differences consist in more than complexity. It is not simply that there are more variables, more relevant things that can go wrong. It is that chains of causality are not like webs of implication. From the point of view of large social-political assemblages, we may still draw a causal line, but the series becomes cumbersome and implausible (e.g., capital → labor → land → unemployment → poverty → trailer parks → hurricane casualties). Another way of putting it is that it is no longer possible to think of *practice* and *accident* as temporally and functionally distinct (Virilio). The practice contains the accident, not simply as a *possibility*—as that which may or may not happen—but fully and completely as *virtuality*. There is an important distinction. Discourses of risk (risk analysis, risk communication) contribute to a profound—and I would add, potentially perilous—misunderstanding: they consign threat (expressed, for example, as a probability) to the realm of the merely possible.

The Plan

In this work I attempt to foster a productive codependence between poststructural and environmental or ecological concerns. I want to acknowledge and engage what I see as a bidirectional challenge that issues simultaneously from modern theory and from environmental thought. The space in between is the place where I have attempted to position this work; it is the place where questions must be posed, relations identified, and new concepts sought. Gilles Deleuze invented a concept that is of use here: the mediator. The mediator is simply a piece of practical advice. It is about how we create concepts—a topic that both Deleuze and Guattari dealt with at length. A mediator is a manner of creating concepts by engaging the relations between ideas or disciplines. There is no point in simply monitoring the movement between separate ideas or inquiries. There is no use in following creative movements that always exist somewhere else. Rather, from the point of view of the mediator, one must attempt to insert a new series in between, in other words, a new series that displaces the authoritative or established discourse within which it develops. This is how one sets out to not just say something differently, but to say something different.

What follows is a work of inquiry and exploration. One way to describe it is that it concerns the breached and overlapping boundaries of the material and the semiotic, two regimes held more or less comfortably apart in most regions of thought today. Another way is that it is an attempt to map some potentials for thinking differently about some regions of living that are particularly resistant to thought.

Very simply, we begin in "Waste" with the lay of the land. What is waste? What are nuclear wastes? Why do these wastes pose a problem? And what is an accident? Is there a distinction between the natural disaster and the technological disaster? What exactly do we mean by nuclear contamination and decay? Why is the government hiding waste in the desert of New Mexico? Each of these provides a piece of the picture I am trying to build of the uniqueness and the virtuality of nuclear and ecological threats. It provides the basics for what happens next.

In "Dangerous Signs," after a brief digression concerning the time capsule as a cultural and epistemological paradigm, I set out as concisely as possible the past, present, and (deep) future of the Waste Isolation Pilot Plant. This figure (which—depending on context—I will call a marker, a monument, a gravestone, a sign) is a plan for the future. And, like plans generally, it is both speculative and prescriptive. It represents a point in time at which precedent-setting decisions are being made that will bind us to the future in an utterly novel way. By no means do I attempt a definitive history. I nonetheless attempt to focus on the key documents produced over the past thirty years and show how the problem of waste burial has been approached conceptually, theoretically, and in fact. It is a story of a government and a bureaucracy that must come to face some of the most difficult philosophical problems imaginable—problems to do with death, with meaning, with time—in their deliberations about the temporality and toxicity of nuclear wastes.

In "Threat and Trauma" I attempt to work through some of the implications of ecological and nuclear threats. I begin by situating the desert monument as a kind of massive disavowal. My claim is that the desert monument provides a way to read the anxiety of a culture. From here I move into an extended discussion of risk and the pathologies of threat. How do we come to talk about threats without unthinkingly reproducing the errors of risk discourse, on one hand, or consigning threat to a retroactivity (i.e., trauma), on the other? The contemporary disaster and accident provide ample material upon which to reveal the discursive poverty around contemporary ecological threats. The theoretical trajectory here is to sort through a certain ambivalence with respect to the language of psychoanalysis (i.e., the real) as the useful way to conceptualize ecological threats. I work toward a theoretical outline of threat as a lively and creative force, in other words, a philosophy of threat based not on the possible but on the virtual.

Waste

Waste

Drift

MODERN WASTES and nuclear wastes in particular are difficult things to contain—conceptually, discursively, and otherwise. One way to think about it: You take out the garbage and leave it on the curb. Sometime later it disappears, presumably taken to a transfer station, a landfill, a dump. Here your garbage is contained and confined and kept safe. In this series time is not a factor. Another way to think about it: You might take your household organic waste and compost it under the rose bushes in the back yard. In this case it's not about containing the waste, and time is precisely the factor. By dispersing the waste in this way an entirely different series is set into motion. A series that requires time in order for decay and decomposition to take place, freeing the organic and inorganic to the whim of the roses. In the case of the garbage, the question of decay is moot; landfill is purely about making sure the waste doesn't leak.

One commentator puts it this way: "Waste is a spatial category; it is produced in place; it is realized only in its materiality." On the other hand, "Decay is a temporal category, it is produced over time, as duration, it is the process of desubstantiation."

> Waste which successfully enters the process of decay is transformed into energy and is dissipated, lost, expended. Decay can only become waste if its processes come to a halt, and it stabilizes long enough to take form.[1]

Two utterly different economies are involved here: one of accumulation and the other of expenditure. From this point of view there is either disposal *or* containment; one or the other, but not both simultaneously. Leakage occurs when material moves across or between the two series, that is, when that which was contained becomes dispersed (e.g., Chernobyl, Bhopal) or when that which was thought to have been dispersed becomes somehow contained or accumulated (e.g., low pH stack emissions reappearing downwind as acidic precipitation, atmospheric CO_2 concentration, Minimata disease). In the case of Chernobyl the containment was breached, resulting (officially) in local and downwind dispersal. In the second case, stack emissions (the so-called super stack in Sudbury, Ontario, is a good example), what was assumed to have been dispersed (SO_2, NO_x, particulates and derivatives) in fact begins to accumulate "unexpectedly" elsewhere in the form of acidified soils and water.

This type of model adequately describes the relations between disposal and containment only on the condition that we are speaking about materials that operate according to the equivalencies: containment = spatial, and disposal = temporal. But, when material starts moving between series, there is a problem. It requires that one start to think differently about the material and about its disposition. A conceptual shift is needed. When your garbage disappears from the curb, as landfill waste, the remainder of domestic consumption, it can be *disposed of*, it can be contained in a space where in a temporal sequence of events of decomposition (aerobic or anaerobic), that waste will decompose. Or at least most of it will. And if there is a loss of containment, if there are leaks in the containment system, those leaks are more or less problems of a technical nature. Solvable, that is, by technical (engineering) means. Or, on the other hand, it could be dispersed (ocean dumping), where similar processes of decomposition, together with homogenizing, entropic forces of dilution will ensue. It is essential to understand that nuclear material fails to conform to the assumptions of this model. Nuclear material can neither be completely accumulated (contained), nor *spent* (disposed). It tends to drift. The duration over which it must be maintained and protected spatially is too long—it seeks its own disposal.

So what happens if we start to think about all of this specifically in terms of nuclear waste? There is by now a great deal of what is generally viewed as Cold War detritus. It is also true that there could be a great deal less than there in fact is. However, in 1977, President Jimmy Carter made one of those critical historical decisions that was probably both fortuitous and disastrous. He disallowed all plutonium reprocessing and recycling on the grounds that a domestic plutonium fuel cycle economy would present a massive security risk. Accordingly, material that would qualify as fuel in, for example, France, has the status of waste in the United States.[2] But in any case, there is a great deal of it. Some of it is piled up, some is partially buried, some is lost, some is leaking through its temporary containment apparatuses. A startling example is the Hanford site in Washington. The intrigue, the cover-ups, the covert experiments on workers and area residents, all of those pale in comparison to the simple fact of the accumulated waste (nuclear and otherwise) that permeates the site. Materials once contained in ponds and other confinement areas have now infused the site to such a degree that the entire area must now be considered a de facto waste repository.[3]

Yet it would be incorrect to think that the question of nuclear wastes today is one pertaining only to industrial-military leftovers. Whether the Cold War is over or has passed into a new phase of media, mediations and simulation notwithstanding, it is also important to realize that the nuclear problem is not simply one of attending to existing accumulations of waste.[4] It's not just about clean-up; there is always the question of whether someone might put it to use. This would be the threat of the so-called dirty weapon (a conventional explosive device surrounded by nuclear waste). Of course this is not really a question at all. It is only by rhetorical flourish that the now common practice of making munitions from depleted uranium (read: waste) is spun differently. In both cases it amounts to a distribution of the effect, without dropping the (big) cause. Something on the order of 860,000 rounds were used in the Gulf War, and 31,000 rounds were used in Kosovo and Albania in 1999. Whether this really is an attempt at an industrial-military plus-sum solution to war and waste, the fact remains that with the present generation of armor-piercing weapons, depleted uranium is a remarkable substance for the manufacture of projectiles. When it finds its target, traveling some five times the speed of sound, the projectile vaporizes in an inferno of burning metal, while liquefying and incinerating the metal it contacts.[5] Apart from critical statements of concern from agencies such as the World Health Organization, there is little agreement about the long-term health implications of these weapons.

In any case, such waste—whether in a leaking drum, spread out over greater metropolitan Baghdad, in Kosovar ruins, or yet to be produced—is no longer thinkable as pollution, as matter out of place.[6] Rather, it must be seen as a novel feature of this point in history. Novel, because it represents a *new form* of waste. It really is matter *without* a place. A kind of waste that resists its own containment. A kind of waste that operates in a radically different temporality; it is material whose toxicity requires a different conception of history and time. The degrees of freedom (i.e., the number of relevant variables that must be taken into account) within which assumptions of containment probability operate in the case of a landfill are of a radically different order from those of nuclear materials. Consider: nickel-59, with a half-life of 80,000 years, will remain radioactive and dangerous for upward of 750,000 years (a conservative estimate, given that the rule of thumb for radioactive abatement is 10 half-lives). Within such temporal limits, probability models of containment failure converge on certainty in an asymptotic manner.

The sort of time and history that must be grappled with in the case of nuclear materials is precisely the challenge that the WIPP marking system would seek to address. The time that must be thought is a discontinuous time. A time in which "our" world can cease to be. Too remote to conceive of as connected to us through relationships of filiation. It is a time that approximates pure future, too distant to seem connected to a present. And equally, it is a time that challenges one's sense of history. To begin with, it challenges history as a record of permanence by casting that very permanence into radical doubt. And the distance that must be conceived of, the utter magnitude of the "future" that must fall under administrative control exceeds the cumulative historical record from which inductive support may be drawn. Ontological anxiety on the one hand, staggering hubris on the other.

Conjunction

Cesium 137 in the fallout, by affecting reproductive cells, will produce some mutations and abnormalities in future generations. This raises a question: are abnormalities harmful? Because abnormalities deviate from the norm, they may be offensive at first sight. But without such abnormal births and such mutations, the human race would not have evolved and we would not be here. Deploring the mutations that may be caused by fallout is somewhat like adopting the policies of the Daughters of the American Revolution, who approve of a past revolution but condemn future reforms. Causes much less involved than radiation have the effect of increasing the number of mutations. One such simple cause is an increase in the temperature of the human reproductive organs. Our custom of dressing men in trousers causes at least a hundred times as many mutations as present fallout levels, but alarmists who say that continued nuclear testing will affect unborn generations have not allowed their concern to urge men into kilts.

Edward Teller, *The Legacy of Hiroshima*[7]

In the infamous essay "The Tragedy of the Commons," Garrett Hardin identified pollution as a member of a set of problems that shared the unique characteristic of having no formal solution. Nature, he said, could be the only arbiter of the negative dynamics resulting from an increasing human population. Nature will commensurate the incommensurable, he said. A shifty rhetorical move, but one that lifted the burden of ethical thought in a manner reminiscent of a Hobbesian nature bats last. But Hardin was right, at least in so far as he pointed to a category of problems that share a characteristic of having no formal solution. It is, however, perilous to think that "nature" will then make the difficult decisions on our behalf. The atomic bomb draws its meaning not from nature, but from the hands of humans. "The fear produced by a tidal wave or a volcano has no meaning." Or so Bataille figured it.[8] Neither makes one afraid in order to make something else happen (surrender, compliance, etc.). This is also true of nuclear wastes but for reasons that are quite different.

Each level of the nuclear waste problem is mediated at another level by other problems and other systems. The drift of nuclear waste from a disposal facility is in one sense conceivable as a purely technical problem of containment design. But this realm is mediated at other levels by legislative design, by risk models, by social perceptions of need, by various ideas of liability and its limits, and so on. Since the formal characteristics of each of these systems are different—presupposing different ideas, different criteria for what would count as evidence—there would seem to be no way to optimize for a solution without having either an enormously elaborate model of the relevant systems and their interaction(s), or—and perhaps in any case—endeavoring to make a viable and working reduction of the complexity involved in order to consider only those interactions felt to be relevant. This would seem to present itself as a problem of optimization, for example, optimizing for social good, economic viability, and maximum containment. However, not all of these systems are equivalent, much less commensurable. Containment must be optimized in and of itself. Yet to do so, the other variables under consideration cannot likewise be optimized.

In this sense, nuclear threat asserts a unique problem; a problem that demands a solution even while admitting of no solution in particular. As problem, it is located imperfectly wherever we might wish to locate it. It is not helpful to simply say that the field of nuclear threat is as much a discursive and epistemological issue as it is a material one. One needn't move this into an argument about social construction. Threat is certainly socially constructed, if what we mean is that it could be different than what it is, but this doesn't get us very far.[9] To understand the virtuality of threat one must also come to terms with the sobering conclusion that *as* virtual, it is also real. It is not a matter of discourse on the one hand, material on the other. It is about conjunction, "and . . . and . . . and . . . ," as Deleuze and Guattari put it. Threat as virtual *and* real, as discursive *and* nondiscursive, as epistemological *and* ontological.

The Nuclear

What does all this mean? It matters. Radioactive materials are simply understood to be seriously dangerous. But to be simply understood can too easily mean *understood simply*. We cannot dispense with a very overt realism when speaking of such things as radioactivity. We take as a matter of faith (an odd faith though it is) the horror of radioactive poisoning, of radioactive death resulting from violent subatomic fracturing of materials, of bodies. We can recall, for example, the terrifying routineness of the mortification of bodies in Imamura's *Black Rain*. Or the traumatic incommensurability of Alain Resnais's *Hiroshima mon amour*. We can say neither that our symbolic, discursive constructions simply *miss* their object, nor that the object itself can be entirely hit. The profoundness of this is easily lost amid the conventional discourses of heroes and villains, winners and losers.

To the extent that North Americans pay attention any more, we are stuck between the radioactive hysteria of the 1950s and 1970s, the soothing words of the nuclear industries, and the promise of desert cemeteries for waste. (Although in the months following September 11, 2001, the possibilities of further adventures in panic seem not far off.) This too may be changing. With the rolling blackouts in California getting hijacked by arguments around deregulation, the nuclear industry is taking advantage of the increasingly diminished half-life of nuclear opposition— "It's clean, it's too cheap to meter"—sound familiar? We perhaps remember that plutonium has a half-life of 24,000 years, though we may not be sure what that exactly means. And we all remember Chernobyl and may indeed have some vague images of reindeers and stock footage of Lapps in Scandinavia as having figured into the story. Indeed in the days and weeks following the events at Chernobyl, reindeer functioned as a provocative Christian/Disney plot device that allowed a complete story to be pulled together from the scarce and contradictory reports being released at the time from the Soviet Union.

The ground zero, as it were, oscillates for most of us between Three Mile Island—*what happened there anyway . . . nearly a meltdown?*—and the bombing of Hiroshima and Nagasaki.[10]

These events, these mishaps, tend to disappear, to become incorporated into other aspects of cultural memory, if at all. For instance, the explosion and coolant release at Chalk River, Ontario (1951): control rods were inadvertently lifted from the core, resulting in a hydrogen explosion and the flooding of the reactor building with nearly a million gallons of highly radioactive water.[11] Or the disastrous reactor fire at Windscale, U.K. (1957): this reactor, designed to produce weapon-grade plutonium, had a maintenance cycle that required the periodic discharge of stored energy. During one such event in 1957, the fuel ignited. The resulting blaze lasted several days and involved the significant discharge of airborne radioactive material.[12] In 1961, during a maintenance routine, a reactor explosion occurred in Idaho Falls that resulted in the immediate death of three workers (one was impaled and left pinned to the ceiling). This accident is presumed to have been the result of control rods having been removed from the reactor core.[13]

Or in some cases, the events disappear without ever having been known. The best (least known) is the still unreported (major) accident in Chelyabinsk in the former Soviet Union that occurred in 1958. Until very recently there have been few reliable reports as to the precise nature of the Chelyabinsk "accident."[14] It is now known that during the late 1940s the Soviets constructed a very large and highly secret complex of reactors—the Mayak Chemical Combine, in the province of Chelyabinsk—bordered by Siberia to the north and the Urals to the west (and rumored to have been the actual surveillance target of Gary Powers's U2 in the 1960s). This region is now believed to have undergone not one, but a series of nuclear accidents. The first occurred over a period of a decade, in which high-level waste from the reactors was discharged directly into the Techa River—the principal water source for several thousand people. The second occurred in 1957, when the cooling system for a high-level waste containment system malfunctioned, overheated, and exploded, exposing over a quarter of a million local inhabitants to a reportedly massive amount of atmospheric radiation. And the third accident occurred in 1967, when Lake Karachay—used since the early 1950s for dumping liquid nuclear wastes—was so severely depleted by a regional drought that sludge dried out, became airborne, and *dust-bowled* an area thought to be over twenty-five thousand square kilometers.[15]

Chernobyl, it seems, has largely disappeared with the Soviet Union.[16] Just as the containment structure of the Chernobyl reactor facility was made transparent by an overheated core, so were the faulty and dysfunctional institutional controls that supported the facility itself. And indeed the latter is about all that has resisted forgetting in popular memory. Chernobyl happened, it seems, not because of an event concerning an insupportable risk, but because of a corrupt and inept political and social configuration that supported it. (The Western media were nearly unanimous on this point, adding for good measure an apocalyptic spin: Chernobyl as "all that is given to us to know the end of the world."[17]) This *post hoc ergo propter hoc* has slipped into Western thought both as alibi and explanation. If we really believed that the reactor was run by "peasants"—and officially sanctioned as such—then the "accident" would seem inevitable for entirely institutional reasons. In a way we are thus empowered to forget *what* happened in favor of remembering only *why*.

Another way to look at this would be to say that even without the Cold War prejudice that constitutes the phantasm of the Soviets as an ultimately backward culture of corruption, we really have no idea what happened because we really don't have the understanding of what such processes as nuclear power generation or fuel production or weapons production involve—or for that matter, what an "accident" is all about. We know it is risky business. We know that accidents can be disastrous. And we know that it is controversial on at least a couple of levels. But beyond this, I think it tends to be a bit of a fog. And rightly so.

Accident

Consider Three Mile Island. The "accident" that occurred there in March 1979 is one of the important reference points in North American nuclear history. It is also an extraordinary testament to the "improbable" and the analytics of the accident.[18]

Dozens of accounts of this event have been written, and I will not attempt to do any more with this example than to show its complexity.[19] The first few minutes of the "incident," as it became known, were something like this: first, the secondary cooling system (the isolated system that transfers heat from the primary cooling system in the reactor core) failed. A system responsible for removing particulates from the secondary cooling water leaked into a nonrelated pneumatic system that controlled instruments (think of this as mechanical synesthesia). The now damp instruments reported a nonexistent error and fed back into a pump-shutdown sequence. Without the pumps, the secondary cooling system was no longer circulating water, resulting in a buildup of heat in both it and the primary system. When the pumps shut down, the turbine that accomplishes heat transfer between systems also shut itself down; without the turbine, there was no way for heat to be released from the core. In such unlikely scenarios, a redundant system exists in order to circulate water through the secondary system, and thus prevent heat buildup in the core. However, the valves that allow water to flow from the emergency reservoir into the secondary system had—for reasons unknown—been left shut. The control panel indicator gauges that would have clearly shown the operators that these valves were in the wrong position were unfortunately obscured by a repair tag hanging on the console. With no heat reduction in the core, the reactor was scrammed (meaning that graphite control rods are dropped into the core to slow the reaction). But even with a slowed reaction, the decay products continued to react, and with no cooling systems operating, the core was still getting hotter. In such instances, a safety valve exists that allows the operators to directly bleed pressure from the reactor vessel. However, when the operators opened this valve and released much of the built-up pressure from the reactor core, the valve failed to reset into the closed position. As a result, about 40 percent of the water from the core was expelled, creating a context for the situation popularly known as the China Syndrome; that is, a meltdown. In fact, what was happening was a loss of containment event; the reactor core was becoming exposed. The operators, however, knew none of this. Nor could they, because on one hand, the instrumentation reported conflicting and nonrelated errors, and on the other, the failure-mode assumptions that they had been trained to make did not include the failure mode they were in fact currently experiencing.[20]

Thus the incident commenced with a series of events that *could not* happen, and, therefore *were not* happening. This is more than saying that the events were unexpected and incomprehensible. The system performed in a way that was outside of the universe of belief of the operators. The instrumentation, assumed to be a reliable index of the reactor's operating envelope, was cut loose and turned into a panel of floating signifiers (Levin) and began to communicate either the wrong information or none at all.

The fact that various warning alarms, buzzers, and a thousand or so warning lights were simultaneously flashing, honking, and buzzing only made the situation more chaotic. Similarly, the fact that when the Three Mile Island management concluded that an "incident" was in progress they were unable to contact the Nuclear Regulatory Commission (a message had to be left with the answering service) slowed response time significantly. Add to this other circumstantial developments—such as the simultaneous failure of independent systems effectively coupling isolated systems and that the site computers had become so overwhelmed generating diagnostic reports that they were unable to prioritize the massive queue of data, delaying the output of important information by hours in some cases—and it is easy to see how the possibility for a decisive response became increasingly remote.

Incomprehensible events persisted for the next day and a half, culminating in a situation that was about as close as you can get to a "worst case" scenario. It was not until ten years later when it became possible to inspect the reactor that it was discovered that some 20 tons of uranium had melted onto the bottom of the reactor vessel.

The accident at Chernobyl was no less incomprehensible. In 1986, when Unit 4 failed and exploded, rendering the core transparent to its environment, there was no objective way at hand with which to characterize the damage. There was no way to assign responsibility. There was no way to characterize a jurisdiction in which a risk might be evaluated. Indeed, the very concept of the casualty had to be reorganized temporally to include the category of those not yet born. There was no way even to properly locate the "event." Indeed, the event could be said to have begun not in the Ukraine on April 26, 1986, but two days later, some 100 or so kilometers north of Stockholm where what was discovered was only a semiotic problem. That is, drifting radioactive signs (indices) were discovered only after a worker triggered a radiation alarm when *arriving* to work at the Forsmark nuclear generating station.

Fatality of Necessity

What precisely can be said to constitute the "accident" at Three Mile Island is not at all clear. To consider the sequence of events as they took place one would conclude that the "accident" was really a kind of utterly improbable series of nonrelated failures that involved electrical, hydraulic, servo-mechanical, computer, administrative, institutional, organizational, interpersonal, and other structural and epistemological factors. It was an assemblage-level failure; that is, the "system" that failed was far larger and more complex than those involved had realized.[21] In one sense this points to the fact that when a complex system such as a reactor moves rapidly away from its "normal" operating envelope it can and will behave in ways that are "incomprehensible." In another sense, it means that the only way to adequately speak of the risks involved would be to take fully into account the social, material, semiotic, and political factors—a daunting if not impossible feat. (Though parenthetically, a task that has spawned the discipline and industry of modeling.)

And there is still another sense—one that really begins to suggest something quite interesting and disturbing about the contemporary accident, about the technoscientific disaster. As Paul Virilio likes to point out, *any* technical production is simultaneously the production of a typical accident. Failure is programmed into the product from the moment of its conception; the ship begets the shipwreck, the train, the rail catastrophe.[22] And yet these are local catastrophes. They happen somewhere to someone and are caused by something. The contemporary technological accident does not follow this specification. To Virilio this suggests that the classical Aristotelian distinction between substance (as the essential, as that which does not exist in another) and accident (as relative and contingent, as that which exists in and is said of another) has come under a reversal. But it is not the specific Aristotelian accidents (that is, quality, quantity, relation, action, passion, time, place, disposition, raiment, and so on) that have ceased being nonessential. (Indeed, it should be clear that at least some of these accidents have become even more inessential.) So when Virilio says that now it is the accident that is essential and substance itself that is relative and contingent, what is meant is that the accident is no longer something that can be added-on (to a technical production) without at the same time radically modifying that substance. Somewhat predating Virilio's thoughts on the matter, Octavio Paz put it this way:

> Axiomatic, deterministic systems have lost their consistency and revealed an inherent defect. But it is not really a defect: it is a property of the system, something that belongs to it as a system. The Accident is not an exception or a sickness of our political regimes; nor is it a correctable defect of our civilization: it is the natural consequence of our science, our politics and our morality. The accident is part of our idea of progress as Zeus's concupiscence and Indra's drunkenness and gluttony were respectively part of the Greek world and of Vedic culture. The difference lies in the fact that Indra could be distracted with a sacrifice of *soma*, but the Accident is incorruptible and unpredictable.[23]

Paz captures something very important about the situation. It's not just that technologies and technical productions contain their own accidents, nor that Aristotelian accidents have become more substantial, but that the accident itself is always a part of our time, our progress. The accident is not accidental; it has become a "paradox of necessity: it possesses the fatality of necessity and at the same time the indetermination of freedom."[24]

Today we may put this differently. It is true that the accident is not accidental. The accident is part of the endeavor. The accident is not the empirical falsification of human endeavor, as some disaster theorists have it. It is also true that the accident's threat is real without being actual. We could then say that threat is the virtual aspect of the accident. Threat is certainly real (but not actual), it is, as Paz said, possessed with the fatality of necessity, *and* the contingency of freedom. Threat always has one foot in a virtual space, with the other in the accident.

Nature and Not

We could ask how different is the technical accident from, for example, an earthquake, or flood, or hurricane. This is after all a comfortable distinction—the natural disaster is in the final instance, natural. Right? The term itself is at once descriptive, explanatory, and normative. But what is this distinction between an accident of a natural sort and one that is anthropogenic or technological? In a way this is an extremely important question. It assumes, on one hand, that there is in fact a meaningful distinction between the natural and technological. There might seem to be an obvious fault line when one thinks of the distinction between, say, flooding and Three Mile Island. In the former there is a kind of punctuated event, a rapid deviation from some equilibrium that begins, eventually crests, then abates. An act of God, or Nature, according to taste—but in any case, *not* from the hands of Humans. Three Mile Island doesn't exactly follow this sequence. Certainly it had a beginning. And certainly it reached its maximum proportions in the weeks following, but it is entirely unclear about how one would place the point at which it ended. Kai Erikson, for example, sees that part of the difference between a toxic event and that of "classical" forms of disaster is in the way that toxic events fail to conform to the rules of plot; that is, the figure of tragedy is itself left incomplete. Beginnings are retroactively constituted. Love Canal is a good example of how toxic disaster begins precisely *because* it really began sometime earlier. And its ending is equally indeterminate. For residents of Love Canal, the events may significantly never end, and for the rest of us, they ended when we forgot to remember them. And of course forgetting to remember is also a salient feature of how it all began; disaster is always implicated in memory.

But this distinction between the technological and natural is only apparently easy to draw. The obviousness of the difference between a flood and an oil spill is only supported by the superficial opposition that the two terms impose. Considered from an assemblage level, the flood may well be as much technological as the spill is natural. And in any case we could say that both would be considered "normal" in the manner proposed by Charles Perrow.[25] The normal accident is inscribed into the design of technological endeavors. When, for example, the safety of a design is given in a proposition such as "The reactor is expected to operate within design expectations x times out of 100 for y hours or years of operation," the subcontrary of the proposition (some S is not P) is also the case. In other words, to speak of a safety probability is to have already inscribed the probability for failure. In this sense the accident at Three Mile Island was normal. There are two reasons (at least) why the same could be said of "natural" events. The first is quite straightforwardly seen in the periodicity of punctuated events: the hundred year storm, the eruption of Mount St. Helens, seismic activity in Southern California, etc. Such events are both knowable and unknowable simultaneously in much the same way as a reactor failure or an oil spill. The second reason is to be found in the general fuzziness of the distinction between the technocultural and the natural. In the case of flooding, one would most certainly have to include in the causal picture many elements that are not at all natural in a conventional sense: patterns of development, deforestation, soil modification, weather patterns, and all the other elements that would constitute the local, regional, and possibly global hydraulic, terrestrial, and atmospheric assemblage. As Erikson put it, albeit more poetically, the collapse of a mine shaft in Appalachia is but the collaboration of a restless mountain and a careless people.[26]

Yet unlike Chernobyl, the accident at Three Mile Island is typically understood as having resulted from operator error. While this has an element of truth, it obviously fails to capture the complexity of the situation. What it does do, though, is cut a political and discursive fault line between the improbable *accident that happened*—in the case of the United States—and *catastrophes of the inevitable*—in the case of the former Soviet Union.[27]

As for the other pole of the nuclear imaginary—Hiroshima and Nagasaki—the fact of what happened there remains a kind of impossible idea for Americans. But perhaps more than any other cultural feature of the nuclear age, these unspeakable events made nuclear threat into a set of images seared into the American consciousness like the shadows of humans scorched onto streets and sidewalks of these Japanese cities. The result was, I think, the onset of both a moral malaise and a nuclear anxiety.[28] Somehow the events became personalized in the sense that the world had changed, a certain innocence was lost, and no one—especially given that those who were killed were simply citizens—no one was safe. But the discourse of a malevolent natural force (so popular at that time) and the harsh reality of threat that it fostered bore witness only to the abstraction of nuclear threat.

In *Nuclear Fear: A History of Images*, Weart gives an interesting analysis of the manner in which the media and political figures spoke of the bombings. He describes how the bombings were framed as the unleashing of nature, that "something unimaginable had come into the World," and according to Churchill, the bombs were "a revelation of the secrets of nature, long mercifully withheld from man." He identifies some compelling social and political aspects to the focus on the "maximum credible accident" scenario (the principal scenario used in regulatory controls). The foremost result of this focus is that little attention is paid to the accidents that have already happened—accidents that were less than the maximum, but entirely (if regrettably) credible. The coupling and complexity type of accident exemplified by Three Mile Island was and has been studied far less than the hypothetical "massive incident." This amounts to attempting to define the "maximum credible accident" as part of the predictable operating envelope, but ignores how chaotic, nonlinear interactions are in fact the norm.[29]

The Strange Alchemy of Decay: Some Things We Should Know About

It comes from outer space, the ground, and even from within our own bodies. Radiation is all around us and has been present since the birth of this planet.

Environmental Protection Agency,
Radiation: Risks and Realities

It seems odd, but there is very little popular or vernacular understanding of the nuclear. I mean two things by this. First, I mean a vernacular understanding that approximates to the discourses of science. And I mean this in comparison to understanding of the body, or medicine, or ecological concepts like that described by ecosystem or bio-accumulation. To put this differently, the point is not just that someone like Rachel Carson has made more of a mark on North American social consciousness than has Rosalie Bertell or Helen Caldicott. There is a very real resistance to constituting a reality to things nuclear—a resistance that can lead as easily to a generalized complacency as to a hysterical frenzy. Nuclear materials ask us to believe in the invisible. So what kind of material is it that engages in what amounts to a nuclear-economic pleasure principle? What kind of material is it that seeks above all a return to the *absolute repose of the inorganic*?[30] What kind of understanding (if at all) is appropriate to the spontaneous material transformation undergone by radioactive elements that results in the emission of radiation?

The landscape of nuclear matter is a strange and foreign terrain. Its principal map, the periodic table—essentially the *Diagnostic and Statistical Manual* of matter—depicts an ontological shoreline that is subject to constant—though, for the last century, predictable—tectonic shifts. Throughout its history it has been in a state of flux: revisions, refinements, additions.[31] The International Union of Pure and Applied Chemistry—the contemporary body charged with the responsibility of legislating matter in and out of existence—oversees the table. Years ago as a student of science I was taught that there were 105 elements. By 1996, there were "officially" 109 elements in the world. A year later there were 112, and now in the early dawn of the millennium there are 118 (the most recent, I believe, being ununoctium).[32]

The vast majority of elemental matter on earth is of a stable configuration.[33] Spontaneous changes and transformations are fortunately the exception. Matter in a stable configuration can be understood as a mingling of (nuclear and electric) forces in equilibrium. The principal atomic constituents subject to these forces are the nucleons of atoms (the most basic of which are protons and neutrons) that collectively constitute the always positively charged atomic nuclei. And surrounding this are the negatively charged electrons. The ratio of charges tends to be at unity.

The forces that are most active within the atomic nuclei are of two sorts: *nuclear* force is a strong force that binds together atomic nucleons and operates only over very small distances and *electrical* forces that, though weaker, operate over greater distance. If one thinks of a chemical reaction, say, burning coal, the kind of rearrangement of matter that takes place—oxygen + carbon → heat energy + carbon dioxide—involves changes at a molecular, but not atomic, level. In other words, a new *chemical arrangement* has been made but the constituent bits of matter are unaltered. The principal forces at play are of a weak electrical nature, and the resultant energy potential is relatively small. However, when reactions take place such that the nuclei of atoms are altered, when the number of protons and neutrons are changed, the resultant release of energy can be staggering, and the matter itself, so to speak, speciates.

The products of such a process of atomic reorganization are of two sorts. The "new" matter is an isotope, and the leftover bits are the decay products. Isotopes can be either stable or unstable. If stable, they are subject to chemical interactions, but not to spontaneous nuclear ones. All elements greater than atomic number 83 (bismuth) are unstable and thus all of their isotopes are unstable as well. The heaviest naturally occurring element—that is, with the largest nucleus—is uranium (atomic number 92). All elements heavier than this must be produced by technical means. And the larger the atomic nucleus, the less stable is the atomic structure, and the more rapid its breakdown. Thus the exceedingly brief existence of ununoctium.

The question of stability of a radioactive element or isotope is a relative one. There are elements that can *exist* for a fraction of a second, and there are elements such as uranium-238 that take 4.5 billion years to decay only partially. The manner in which this is conceptualized is the "half life," an approximation of the length of time it takes for *half* of a sample of unstable atoms to decay. Much like the LD_{50} concept in biological science, it is a statistical concept applicable to aggregates only. The products of decay, the bits that are ejected from the nuclei of unstable atoms, are the remainder of this process in which matter seeks stability. What makes the remainder of this process dangerous, what in other words constitutes the threat of this matter, is its potential to ionize; which is to say, strip electrons from atoms or molecules that it encounters. It is precisely this ability to ionize that describes the manner in which the remainders of decay can cause carcinogenic, mutagenic, and teratogenic changes to living tissue.

Lively Fictions

The most massive and densely ionizing of decay products is the alpha (α) particle. It is a manner of decay that allows an unstable atom to rid itself of excess protons. A single α particle is comprised of two protons, two neutrons, various subatomic denizens, and a single pathological drive: to lose its double positive charge through the appropriation of otherwise engaged electrons. Helium envy: the α particle is simply a helium nucleus minus the electrons. Traveling at some 10,000 miles per second, the α particle—even though it can travel only a fraction of a millimeter within organic material—crashes into one hundred thousand or so atoms before arriving *home*, as it were, in a stable helium configuration. Damage done as a result of the forced ionization is dependent on the particular tissue(s) involved.

The beta (β) particle, another product of nuclear decay, is composed of a single electron jettisoned at 150,000 miles per second from the nucleus of nuclear material in the process of decay. As an electron, it carries a negative charge. The β particle is as close to nuclear alchemy as I can imagine. It poses the question, How does a nucleus expel an electron (since electrons are found outside of the nucleus)? The explanation is apparently that a neutron *just spontaneously* transforms into a proton and an electron, keeps the proton, and jettisons the electron. In any case, β particles, like their α kin, ionize whatever organic material they encounter. Since both the mass and charge of the β particle are less than that of the α particle, its behavior is different. It is capable of much less ionizing. However, because of its lower mass the distance it travels before it regains a stable configuration with a positive ion is at least an order of magnitude greater (i.e., up to half an inch of tissue). In other words, it can penetrate further, but will lead to less ionizing damage.

The third principle mode of decay, gamma (γ) radiation, is nonparticulate. It is purely high-energy photons that are produced in the process of nuclear decay. As photons, γ radiation has no mass and no charge and thus has much greater penetration potential than either α or β particles. The damage wrought by γ radiation is as a result of its action *on* particles. The high energy of γ radiation can displace cellular electrons, effectively creating β particles (and other masses such as positrons) which in turn can do ionizing damage to adjacent tissue and cells.

The manner in which the energy of decay products and their remainders is quantified oscillates around a kind of subject/object split. There are scales that are concerned solely with the number of disintegrations a particular sample of material will undergo in a given period of time. The curie, or in more contemporary parlance, the becquerel, are such scales. The concern is the radioactive object and not the body. The potential is only its activity, the rate at which a radioactive material decays. In order to speak about the effect that radioactive decay has on a body, one must know more than the disintegration rate. In addition, one must know the *kind* of disintegration and its energy level. The units typically used to describe the rate at which tissue absorbs radiation are rads (radiation absorbed dose)—that refer *only* to the extent of tissue absorption (per gram) of energy deposited by various high-speed particles and γ rays. In this sense, rads and grays (100 rads) are significantly less assumption-bound than units such as the rem. The rad and the roentgen are similar in this respect. Rems (roentgen equivalent man) relate the absorbed dose to the effective biological damage in living tissue. But in order to do so, one must have some conviction with respect to the probabilistic basis for various sorts of tissue sustaining damage from certain levels of energy absorption. In other words, units such as rems or sieverts (100 rems) or effective dose equivalent are, in addition to an expression of a quantity of energy absorbed, an expression of the assumed effect on tissue.

All of this atomic activity, even as one attempts the routineness of its language, is just too small, too strange to be more than an article of faith. *So put it in the ground and be done with it.* In the absence of a direct experience, the science fiction of decay inhabits the liminal regions of perception and belief, the phenomenological and the psychological. Its existence must be granted and agreed on by purely symbolic means, and yet the stakes are surely more than symbolic. The question here is not *what is the virtual?* but *where* (or how) *does the virtual intervene?* How are its actualizations felt in the world? Radioactive objects are not simply the fictive productions of a working scientific paradigm; they both can and have been put to the test. Blowing the whistle on the ideological or political affinities of science is not of much use, in this respect at least. This is more than a temporary victory in a game between winners and losers in the production of scientific truths. One must come to accept that they are both the fictive productions and the absolute creative and real achievements of contemporary scientific practice. They are both. They are both even as we wait patiently for the next Copernican revolution in the construction of matter. And, and, and.

We can say this: threat makes things happen. Or, perhaps we could put this differently: nuclear materials stand in relation to their containment only very imperfectly—there is always leakage. There is leakage too when one thinks of nuclear threats as virtual. Conceptually we can understand the virtual as real. But this may seem *only* a conceptual operation. It is necessary to see as well the difference that this kind of understanding can make (that is, nonconceptual difference). But we have no real *experience* of the virtual of threat apart from its actualizations—this is its leakage. The virtual leaks actualizations, not all of which are accidents: how threat is organized, how meanings are assigned to it, and how these meanings are then endowed with the capacity to carry out social work and to undertake translations between the identified threat and (distant) social realms.

The Cold War years are a veritable theme park for looking at translations and transformations brought about by threat. In the cold heat of the duck and cover years, the responsible citizen was required to submit to a series of civilian defense strategies—nuclear fictions par excellence—in the name of the threat of a nuclear war.[34] As Diefenbaker put it in 1961:

> Notwithstanding what has been and is being done, nuclear war is possible either by the intended actions of evil madmen or by miscalculation...your personal survival can depend upon you following the advice that is given and the survival of many others may depend on how well you have heeded the advice.[35]

This is disingenuous, really. He leaves out the other possibility, the possibility that betrays the omnipresence of threat: Us. Therefore, and retroactively, the formula becomes: nuclear war is possible by either "us," by "them," or by mistake. That pretty much covers all the bases. Hence, civil defense. The home fallout shelter. Provisions. Education: duck and cover. And most of all, there was the instilling into social consciousness of the need, the responsibility, to be always alert. More than the need to be prepared, there was the need to be on guard against the unseen threat of nuclear terror. But this was a long time ago. And today? The omnipresent doomsday clock of the *Bulletin of the Atomic Scientists* ticking ever nearer to the midnight of civilization; the metronome of threat.

The Site Must be Marked

The disposal of radioactive waste is an international problem, and although present political boundaries shape many aspects of how the problem is being defined and handled today, it is clear that these boundaries have no relevance to the generations of future millennia. *It is therefore essential that any WIPP markers be designed as part of a global system of marked sites.*

Kathleen M. Trauth, Stephen C. Hora, and Robert V. Guzowski, *Expert Judgment on Markers to Deter Inadvertent Human Intrusion into the Waste Isolation Pilot Plant*

In global terms, the United States houses about one quarter of the accumulated global store of nuclear wastes (estimated to be on the order of 36 million cubic meters),[36] and other nuclear nations are watching the United States to see what manner of success will be had.[37] The Department of Energy (DOE) is the agency responsible for the now decaying domestic nuclear infrastructure. Created during the Carter administration from its predecessor, the Atomic Energy Commission (1945), the DOE is a nightmarishly vast (over 100,000 employees), virtually self-regulating (that is, not overseen by the Nuclear Regulatory Commission), staggeringly expensive (budget for 2000: $17.4 billion), and powerful bureaucratic structure. In addition to some three quarters of a million metric tons of nuclear materials (depleted uranium, fuel-grade plutonium, thorium, and uranium), it controls nearly 2.5 million acres of real estate holdings, electrical distribution systems, 1600 research laboratories, 89 nuclear reactors, and 665 production facilities, more or less.[38]

At present, more than thirty countries operate nuclear reactors for energy purposes. This, together with the eleven countries known or suspected to be operating reactors for the purpose of weapons production, is sobering. More sobering still is that worldwide, including the eight officially acknowledged nuclear states, there are more than 35,000 nuclear weapons. Most of these—the vast majority—are in the United States and Russia. And more than half are operational, ready to go.

To date, the WIPP is the only permanent disposal site in operation. The proposed commercial, high-level waste repository, Yucca Mountain in Nevada, may never actually open. As of February 2002 President Bush has given the Yucca project his support and presumably sufficient funding to make the site operational, but it must still pass through a number of congressional, state, and regional hoops. To date it is mired under a landslide of opposition.

Today, the high-level waste initiative that appears to be moving most rapidly toward realization is in fact a private venture, which, predictably, circumvents a good deal of regulatory machinery. It is made up of eight nuclear utilities— Xcel Energy, Genoa Fuel Tech, American Electric Power, Southern California Edison, Southern Nuclear Company, First Energy, Entergy, and Florida Power and Light— under the name Private Fuel Storage (PFS).

Since 1996, PFS has been aggressively courting the Skull Valley Band of Goshutes to use their reservation located west of Salt Lake City as a location for the "temporary" storage of commercial irradiated fuel. In order for the PFS facility to go forward, the Nuclear Regulatory Commission must approve a twenty-year license, the Bureau of Indian Affairs must approve a twenty-five-year lease for the land, and the Bureau of Land Management must approve a right-of-way to construct a rail line to the site. PFS intends on securing these approvals by early 2005 and now awaits approval from the Atomic Safety and Licensing Board. From the point of view of the Skull Valley Band, the storage of 40,000 tons of high-level waste is really only in keeping with the predominant land use in the area. From the Band's Web site:

> South of Skull Valley on traditional Goshute territory there was wild game which roamed the country freely and served as a vital food supply. This area is now the location of Dugway Proving Grounds where the United States government developed and tested chemical and biological weapons. In 1968, chemical agents escaped from Dugway and approximately 6,000 sheep and other animals died. At least 1,600 of those contaminated sheep were buried on the Reservation by the Government. . . .
>
> East of Skull Valley in the area known as Rush Valley there was native sagebrush, pine trees, food plants, and also wild game. Today, this area serves as a nerve gas storage facility for the United States government. The world's largest nerve gas incinerator has recently been built to destroy thousands of tons of these deadly chemicals.
>
> South of Skull Valley lies the Intermountain Power Project which provides coal-fired electrical power primarily for California. Air pollution fills the skies of the Western Desert and impacts the Skull Valley Reservation.
>
> Northwest of Skull Valley is the Envirocare Low-Level Radioactive Disposal Site which buries [low-level] radioactive waste for the entire country. Within this immediate area there are also two hazardous waste incinerators and one hazardous waste landfill.
>
> Finally, north of the Reservation is the Magnesium Corporation plant, a large magnesium production plant which has been identified by the US EPA as the most polluting plant of its kind in the United States. Chlorine gas releases from MagCorp also impact the Skull Valley Reservation. In the citing [sic] of these facilities on the aboriginal territory of the Goshutes, the Skull Valley Tribal Government and people were never once consulted.
>
> In view of the current hazardous waste facilities and nerve gas incinerators surrounding the Skull Valley Reservation, the Band has carefully considered a variety of economic ventures, including the storage of spent nuclear fuel. After careful consideration, the Skull Valley Band of Goshutes have leased land to a private group of electrical utilities for the temporary storage of 40,000 metric tons of spent nuclear fuel.[39]

Perverse as it seems, this makes the WIPP proposal look reasonable.

In any case, this is part of the contemporary backdrop against which the WIPP is taking place. The particular story of the WIPP marker begins in

1979 when by act of Public Law 96–164, the Waste Isolation Pilot Plant (WIPP) was authorized as a research and development facility to

> demonstrate the safe disposal of radioactive wastes resulting from the defense activities and programs of the United States exempted from regulation by the Nuclear Regulatory Commission [i.e., nonenergy related nuclear wastes].[40]

The technical standards of the WIPP were subsequently set out in the Code of Federal Regulations, Title 40—Protection of the Environment, Chapter 1—Environmental Protection Agency; Subchapter F—Radiation Protection Programs; Part 191, otherwise known as 40 CFR 191.

These regulations state that:

> (a)   Active institutional controls over disposal sites should be maintained for as long a period of time as is practicable after disposal; however, performance assessments that assess isolation of the wastes from the accessible environment shall not consider any contributions from active institutional controls for more than 100 years after disposal.

Attempts to secure the site through active means—guards, gates, fences, etc.—are not assumed to be effective for more than 100 years following closure of the repository. By limiting specific future responsibilities around the site, this regulation resulted in the need for the site itself to be designed as a warning and explanation.

> (b)   Disposal systems shall be monitored after disposal to detect substantial and detrimental deviations from expected performance. This monitoring shall be done with techniques that do not jeopardize the isolation of the wastes and shall be conducted until there are no significant concerns to be addressed by further monitoring.

This regulation is a bit of a mystery. Since none of the material to be interred in the site is retrievable, and since the repository rooms ("panels") are designed to be sealed, it is hard to understand how monitoring could possibly be done without jeopardizing the isolation of the wastes.

> (c)   Disposal sites shall be designated by the most permanent markers, records, and other passive institutional controls practicable to indicate the dangers of the wastes and their location.

Here is the substantial piece of the regulation that provides for the desert monument. The passive institutional controls referred to here are such things as markers, public records and archives, ownership and regulation of disposal lands, and "other methods of preserving knowledge about the location, design and contents of a disposal system." The "probabilistically-based" performance assessment criteria revolve around the use of techniques for working with the subjective analysis of experts.[41] The question of the permanence of permanence has periodically returned to inform policy making and interpretation. The concepts of "most permanent markers" has been a shifting terrain. Initially this meant 10,000 years, on the assumption that if something could last and function for this duration it would probably last longer. This is really the only way to make sense of 10,000 years in relation to the quarter of a million years reasonably required to secure plutonium-based waste. In other words 10,000 years is a model. The demonstration of permanence, however, is assumed to be a fraction of the 10,000 year fraction; that is, another model.

> (d) Disposal systems shall use different types of barriers to isolate the wastes from the accessible environment. Both engineered and natural barriers shall be included.
>
> (e) Places where there has been mining for resources, or where there is a reasonable expectation of exploration for scarce or easily accessible resources, or where there is a significant concentration of any material that is not widely available from other sources, should be avoided in selecting disposal sites. Resources to be considered shall include minerals, petroleum or natural gas, valuable geologic formations, and ground waters that are either irreplaceable because there is no reasonable alternative source of drinking water available for substantial populations or that are vital to the preservation of unique and sensitive ecosystems. Such places shall not be used for disposal of the wastes covered by this part unless the favorable characteristics of such places compensate for their greater likelihood of being disturbed in the future. Disposal systems shall be selected so that removal of most of the wastes is not precluded for a reasonable period of time after disposal.[42]

Almost every requirement of (e) is invalidated with the site and placement of the WIPP. The only "favorable characteristic" which might be said to compensate for the high concentration of mineral (potash), petroleum, gas, and other forms of resource extraction in the area, or (depending upon one's interpretation of "reasonable period of time") for the apparent structural instability of the repository rooms, is the millions of dollars that will be saved by not selecting a different site.

Something Must Leak from the Box

The burial project now being undertaken in the desert of New Mexico is such a complex and convoluted undertaking that it is hard to know where to focus. Take the desert, for example. A rich trope in the American frontier imaginary. What is it that makes the desert so interesting? A Euclidean wilderness? Here is what one French cartographer of the deep American desert has said:

> Desert: luminous, fossilized network of inhuman intelligence, of a radical indifference—the indifference not merely of the sky, but of the geological undulations, where the metaphysical passions of space and time alone crystallize. Here the terms of desire are turned upside down each day, and night annihilates them. But wait for the dawn to rise, with the awakening of the fossil sounds, the animal silence.[43]

This desert of Baudrillard's, reading as it does like Rimbaud on holiday *(from the bituminous desert . . . with sheets of fog spread in frightful bands across the sky)*, is a very serious place. For Baudrillard, it is a place filled with signs; it captures the future of the social—its catastrophe—in the utter indifference of geology. Such a place is for Baudrillard merely and significantly a form; the desert organizes features of surface, and superficiality. Why, he ponders, are deserts so fascinating?

> It is because you are delivered from all depth there—a brilliant, mobile, superficial neutrality, a challenge to meaning and profundity, a challenge to nature and culture, an outer hyperspace, with no origin, no reference points.[44]

It is indeed a tempting leap from the sand and the dunes of the desert to the Sands and the Dunes of Las Vegas. But the desert I am thinking of is definitely *more* than a metaphor. To see its shifting, flat, and mobile brilliance one must get out of one's vehicle; a move one suspects Baudrillard never attempted. Baudrillard, the Desert Rat (as Genosko affectionately calls him), was after all only a tourist.[45]

Nonetheless, the desert I am interested in is a desert filled with spatial and temporal depth. It is an ecological and ontological space. A tricky place, brimming full of emptiness. This is a desert that seems to belie Baudrillard's claim of "no monuments, no depth."[46] For in this desert there will be a monument, and there will be unfathomable depth. This desert that I am interested in will be a place that will house an enormous secret; a secret that must be kept and always disclosed—simultaneously.

In a way this is a sure bet. That is, there is a sense in which this requirement for secrecy and disclosure will most certainly be met—for several reasons. The first concerns the fact that this is a project about limits. Everything about the WIPP and the monument operates in a complex relation to a limit. At the limit of civilization; its place is the desert—the other American wilderness. At the limit of history; its time is the deep future. At the limit of meaning; its witness is unknown, abstract, and indeterminate. At the limit of the symbolic; auguring the *language* of the future is a dizzying confrontation with the aporias that obtain when one steps outside of the frame of the present. At the limit of technology; the ability to engineer materials for this unprecedented duration is and remains hypothetical at best. Distributed about these limit regions, failure and certainty are asymptotically related (and the wager is that the oracular abilities of technoscientific expertise may correctly divine the intersection of the two series).

Yet these things constitute only one dimension of the layers of concealment involved. The other dimension (or "another"—there may be more) has to do with what we might call the epistemological anchoring of the monument. Secrets as well, these constitute the anchoring of meanings—produced through reversal and inversion—whereby certain fundamental discursive concepts are covertly recast to accord with the burial endeavor. For example, *responsibility* is shifted away from its ideological affinity for the "individual"—particularly the living individual—and its autonomy, toward the future that must be alerted to the presence of the interred waste. Likewise, *justice* must proceed not from the rights of the living individual, but from the distributive rights of future persons—that is, the category of persons not yet living. And *national safety*, and its institutional spokesperson, risk analysis, shift from the cold war strategy of being as dangerous as possible in order to remain safe, to its opposite: become as safe as possible in order to preserve the possibility of being dangerous. It is far more complicated than this, but I wish to convey here only the general structure of secrecy: both inside and outside of discourse, both inside and outside of the monument, the project in the desert is prefaced by and organized under a structural secret.

The second sense (in which the requirement for secrecy and disclosure will be met), which in a way preempts the first, is that secrets are just like that; that is, they tend to secrete. Occasional theorists of the secret, Deleuze and Guattari have put it this way:

> The secret has a privileged, but quite variable relation to perception and the imperceptible. The secret relates first of all to certain contents. The content

is *too* big for its form . . . or else the contents themselves have a form, but that form is covered, doubled, or replaced by a simple container, envelope, or box whose role is to suppress formal relations.[47]

And furthermore:

These are contents it has been judged fitting to isolate or disguise for various reasons. Drawing up a list of these reasons (shame, treasure, divinity, etc.) has limited value as long as the secret is opposed to its discovery as in a binary machine having only two terms, the secret and disclosure, the secret and desecration.[48]

Thus it is only as an anecdotal formulation that the disclosure of the secret is its opposite. From the point of view of the concept, however, the perception of the secret is part of it. To paraphrase Deleuze and Guattari, the secret must move through society as a fish through water, but on the condition that society behave toward the secret as water to fish. The secret is a social function, a social assemblage.

Contents

This is the institutional figure of the secret in the desert: a material with contents too big for its form. It is a container of secrets that exceeds itself. It is in this sense that the "significant" part of nuclear material is its own remainder. And the American strategy has been to manage the material (not the remainder) through concealment and selective disclosure. (But I would insist that the concept here of remainder must be understood as not something produced as externality—as with the conventional understanding of industrial effluent, unemployment, etc.—but rather as an excess that is necessarily part of it.) From the point of view of the American government, one could say that rather than one, there are two secrets. On one hand the burial, and on the other the sign. The burial of the waste operating as the justification for the design and placement of the marker. The marker operating as the (ethical) alibi for the interment of the waste. Both concealing a secret operation, and each operating as the standard-bearer for the other. The marker will operate through the deployment of "enduring signs of danger" to signify the danger below. Yet, as we will see, the precise nature of the danger is incidental to the intention of the sign. The signification of this sign—which from this point of view would be the relation between a form and substance of a contents (the waste and its emanations) and a form and substance of expression (the monument as a system of communication and containment)—must exceed in every imaginable way the simple idiom of the monument.

The important thing is that there is no way to reduce the monument to a simple relationship between a signifier and a signified, recto-verso, where materiality is on one side only. The secret operates the way it does because the sign is already doubled—there is substance and form on *both* sides. Nonetheless, the sign's real function is to efface the burial—this is the other secret of the sign. The sign's double mission: efface the waste and remain "dangerous." And the burial, ostensibly the thing that supports the sign, is allowed to remain in secrecy. (One only has to start inverting some of the presuppositions behind these operations to see how, at this level at least, the whole thing might fail.)

The secreted (radioactive) materials, part of the contents in this case, have a very slippery relationship to their form. In a sense the materials themselves are not dangerous; it's what is expelled that is. On one hand, without the particularity of the contents as substance, the material could not have the actual form that it has, but on the other, the form of the contents is only probabilistic (i.e., the decay series of half-lives), related back to the state of the contents at a given moment; an arrangement that can only be known through some kind of disclosure or leak.

> The secret has a way of spreading that is in turn shrouded in secrecy. The secret as secretion. The secret must sneak, insert, or introduce itself into the arena of public forms: it must pressure them and prod known subjects into action.[49]

Thus, secrets can never be perfectly secretive, can never *win* the struggle against disclosure, for they are not in opposition *to* it.

Consider the example of the stealth aircraft mentioned by Baudrillard. A stealth aircraft is paradigmatically a contents that presents *no* form (the precise opposite of a decoy)—this is what allows them to remain unseen. Indeed, as Baudrillard points out, early versions of these aircraft were so transparent, so invisible that they were unable to locate even themselves (resulting in several rather unfortunate and expensive crashes). These prototypes were too secretive. The problem is that there must be some relationship to perception, to the perceptible—"something must ooze from the box," say Deleuze and Guattari. Or, from Baudrillard's perspective, "As is well known, when playing hide-and-seek, you should never make yourself too invisible, or the others will forget about you."[50] And this, he surmises, is the reason why the stealth—even though it was a "high level" secret—was presented to the public to begin with.[51]

Obviously, the secret exists in a relation to the visible and not just to perception per se. The danger of the waste, although not visible, is not even on the order of the visible as such. (And this, I would add, is critical to all thought about ecological and nuclear threats.) Yet the materials to be interred, the materials that emit the danger, are precisely on the order of the visible, and this is why they are to be concealed beneath the desert. Thus there are two sorts of relation to secrecy: one on the order of an invisible visible—the concealed, the buried, the stealth, hide and seek—and the other on the order of an absolute invisibility—radioactive emanations and ecological threats generally (to which I will return later).[52]

In the latter case, the nuclear toxicity operates necessarily outside of the register of sight. It requires a mediation in order to be disclosed. That is, a relay into a signifying semiotic regime: a Geiger counter to render it sonorous or audible, or a body or tissue to be transformed by the absolute invisibility of α and β particles and γ radiation; in other words, nothing renders it visible, it can only relay into the order of the visible via the production of signs (signals, sounds, symptoms). In the former sense, that of an invisible visible, the transuranic material, the dross, whether beneath the ground or heaped on the surface, always contains within it another sort of secret (the significant part of its story left untold). A secret of a different order, and one disclosed by very different means.

In the initial planning phases of the WIPP project in the early 1980s, a null hypothesis—the option of not marking the waste at all—was given some serious consideration. The idea was that if it were really well hidden, and hidden in a place that no one would ever think of looking for it or anything else, then the safety of the present and the future would be secured.[53] By virtue of 40 CFR 191 (see above) however, the possibility that a disposal site could be designed without a permanent marker system was specifically excluded by the Environmental Protection Agency ("Disposal sites shall be designated by the most permanent markers"). If, in other words, the wastes were hidden too well, one might simply forget that they were there and discover either it or its secretions by "accident."

Dangerous Signs

Dangerous Signs

A Digression

Rosetta Envy

PLUCKED FROM the Western delta of the River Nile at the end of the eighteenth century, the Rosetta stone is the structural model of time capsule wish fulfillment. That is, it is both a wish to be understood by the future and an acknowledgment of the incomprehensibility of the past.[1] More recently, the time capsule that was buried during the 1964 World's Fair bore a stainless steel plate bearing an inscription in the then official languages of the United Nations (Chinese, English, French, Russian, and Spanish). But that one wasn't meant for us. Others, though, are. As we have crossed over into the twenty-first century we have become the intended recipient of many time capsules addressed to us c/o the year 2000. It was once a very popular cultural pastime; a message in a bottle from the civilized world into a future unsure. And even today—perhaps *particularly* today—one can buy a wide selection of time capsule kits from places such as Future Packaging and Preservation in Covina, California. Another company, LegaSEA, will sell you a large glass orb into which your ashes or anything that fits may be placed and then consigned to the depths during an expensive day-long, catered cruise. For those with a more messianic bent, the orb may be configured to break free of its anchor, rise to the surface, and presumably wash ashore at some specified time in the future (50, 100, and 500 year interments are available). For those wishing a more permanent arrangement, orbs may be configured to sink to a specified depth, and there ride the subsurface currents indefinitely.

Arguably the most elaborate time capsule message (also not addressed to us) was the project to send an interstellar record with the Voyager spacecraft; the two nearly identical craft were launched in 1977. These spacecraft were to collect and transmit images back to Earth of the outermost planets. Once these vessels completed their work in our solar system, they would simply carry on into space as the third and fourth human artifacts to escape entirely from the solar system. Pioneers 10 and 11 (1971 and 1972), which preceded Voyager in outstripping the gravitational attraction of the Sun, both carried small metal plaques identifying their time and place of origin for the benefit of any other spacefarers that might find them in the distant future. For the Voyager missions, NASA was persuaded to develop a much more ambitious message—a time capsule, intended to communicate a story of *our* world to extraterrestrials.

The Voyager message is basically written on a phonographic record—a 12-inch gold-plated copper disk containing sounds and images. Each record was encased in an aluminum jacket, together with a cartridge and needle. In addition, there are instructions, in symbolic language, explaining the purpose of the spacecraft, its origin, and just in case, it also indicates *how* the record is to be played; for instance, that it is partially audio, and that it is to be played at exactly 16 2/3 revolutions per minute. Since this was the time of a growing interest in extraterrestrial communication—the Search for Extraterrestrial Intelligence (SETI) project having recently begun—all space probes that were launched with trajectories that would exit our system were seen as potential extraterrestrial greeting cards. Carl Sagan was asked to design a "message" for the outside of these vessels. This was not the first time Sagan had been asked to design a message for distant others. In 1974 he designed a plaque for the Laser Geodynamic Satellite. This very high, very stable satellite was designed to orbit the earth for 8 to 10 million years, allowing laser tracking of continental movements. As Sagan summed it up, "This is sufficiently far in our future that a great deal of information may be lost between now and then."[2] (One would surmise that this is a safe bet.) For the Voyager project he assembled a team of scientific overachievers, and an extraordinary project ensued, although within the astronomical scientific community his project was contentious. Some saw it as a mobile and preemptive greeting, of advancing evidence of our intelligence to *like-minded* others. Lewis Thomas, for example, is said to have wanted to send the complete works of J. S. Bach, but then realized that "that would be boasting." British astronomer and

```
1111000010100100001100100000000100000101001000001100101100111100000110
0001101000000001000001000010000100010101000010000000000000000000010001
0000000000101100000000000000000000001000111011010110101000000000000000000
0010010000111010101010000000001010101010000000000111010101011101011000
0000100000000000000010000000000000010001001111110000011101000000010110
0000111000000010000000001000000000100000001111100000010110010111011010000
0000110010111110101111100010011111100100000000000011111000000101100001111
1111000001000001100000110000100001100000001100010100100011111001011111
```

Figure 2. Message of 551 characters.

Nobel laureate Martin Ryle (nephew of Gilbert) actively, though unsuccessfully, attempted to have the International Astronomical Union vote a resolution—in the interest of the safety of the Earth from malevolent others—to the effect that no such messages should *ever* be sent. Once the Voyager spacecraft leave our solar system they will find themselves in empty space. It will be forty thousand years or so before they come within even a light year of the star AC+79 3888 and millions of years before either might make a close approach to any other planetary system. As Sagan noted, "The spacecraft will be encountered and the record played only if there are advanced spacefaring civilizations in interstellar space. But the launching of this bottle into the cosmic ocean says something very hopeful about life on this planet."[3]

Undeterred by any suggestion that the interstellar greeting was anything other than a necessary testimony to the uniqueness of humanity, Sagan and his colleagues were able to design a most amazing recording.

In thinking about the problems of transmission and otherness, I conducted a small experiment in the winter of 1995. I sent twelve friends an e-mail message containing an encoded *signal* similar to that sent on the Voyager. The message contained a short introduction explaining the problem and was followed by a string of 551 zeros and ones, as in Figure 2.

To "decode" this properly required a number of steps. First, one might have recognized that the string of 551 characters is nothing as exotic as a Fibonacci series or even a really big prime. It is, however, the product of two primes, and this turns out to be quite important. In any case, one would have to *see* that the data *wanted to be* information. That is, that the string of zeros and ones had a potential to *mean* something. Obviously this is a critical aspect of the WIPP monument design as well—the monument must assert more than its presence. It must also assert that it is significant beyond its perceptible form. It must assert depth.

In the experiment that I conducted, one would have to see that the numbers both could and ought to be arranged in rows. But not just any rows. They needed to be arranged in twenty-nine rows of nineteen characters each; that is,

```
1111000010100100001
1001000000010000010
1001000011001011001
1111000011000011101
0000000010000010000
1000100010101000001
0000000000000000000
1000100000000001011
0000000000000000000
1000111011010110101
0000000000000000000
1001000011101010101
0000000001010101010
0000000011101010101
1101011000000010000
0000000000000100000
0000000010001001111
1100000111010000010
1100000111000000010
0000000010000000010
0000001111100000010
1100010111010000000
1100101111101011111
0001001111100100000
0000001111100000010
1100011111110000010
0000110000011000010
0001100000001100010
1001000111100101111
```

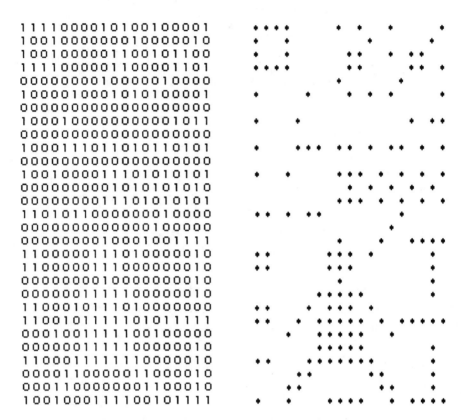

Figure 3. Matrix of twenty-nine rows.   Figure 4. Clarified matrix.

the two primes by which the total is divisible. Then, having made these harmonic leaps of cleverness, one would have to recognize that the rows thus arranged formed a matrix and that the matrix itself was meant to picture. So one would have to shift from making a numerical manipulation (the matrix), to "seeing" the whole matrix as a "picture" with the ones as the bits containing information, representing the figure, and the zeros representing the ground (Figure 3).

Here the metaphor that is most obvious (although hardly universal) would be a television signal. That is, the boundary of the matrix as the boundary of the screen and the numbers as the pixels, either on or off (Figure 4).

Thus visualized, the "picture" contained an image of a bilaterally symmetrical being, the configuration of our solar system, a "likeness" of carbon and oxygen atoms, and the numbers one through five in binary notation. Apparently not many of Sagan's friends could solve a puzzle like this one, though they felt that (other) intelligent life could.[4]

Picture This

The "message" they sent into space was rather large. Try to picture these things: A calibration circle and a solar location map with parameters.

Some mathematical and physical unit definitions. The Sun. The solar spectrum. The planets Mercury, Mars, Jupiter, and Earth.

A map of Egypt, the Red Sea, the Sinai Peninsula, and the Nile. A number of chemical definitions. The DNA structure and the DNA structure magnified. Images of cells and cell division. Schematics of anatomy and human sex organs. A diagram of conception. A fertilized ovum and a fetus. A diagram of a male and a female. A nursing mother. A Malaysian father and daughter. A group of children. A diagram of family ages. A family portrait. A diagram of continental drift. The structure of Earth. An image of Heron Island in the Great Barrier Reef of Australia. A picture of the seashore. A picture of Snake River, the Grand Tetons, and sand dunes. A picture of Monument Valley. A forest scene with mushrooms. A growing leaf and some fallen leaves.

A sequoia. A snowflake. A tree with daffodils. A flying insect with flowers. A diagram of vertebrate evolution. A seashell (Xancidae). Some dolphins. A school of fish. A tree toad. A crocodile. An eagle. Jane Goodall with chimps. A sketch of some Bushmen and Bushmen hunters. A man from Guatemala. A dancer from Bali. Some Andean girls. A Thailand craftsman. An elephant. An old Turkish man with beard and glasses. An old man with a dog. A mountain climber. Cathy Rigby. Some sprinters. A schoolroom. Some children with a globe. A cotton harvest. A grape picker. A supermarket. An underwater scene with a diver and fish. A fishing boat with nets. A fish being cooked. A Chinese dinner party. A demonstration of licking, eating, and drinking. The Great Wall of China. An African house being constructed. An Amish construction scene. An African house. A New England house. A modern house in Cloudcroft, New Mexico. The interior of a house with an artist and a fire. The Taj Mahal. The English city of Oxford. Boston. The UN building and the same building at night. The Sydney Opera House. An artisan with a drill. A factory interior. A museum. An X-ray of a hand. A woman with a microscope. A street scene in Pakistan. Rush hour traffic in India. A modern highway in Ithaca, New York. The Golden Gate Bridge. A train. An airplane in flight. The airport in Toronto. An Antarctic expedition. A radio telescope in Westerbork, Netherlands. The radio telescope at Arecibo. A page from Newton's *System of the World*. An astronaut in space. The Titan Centaur launch. A sunset with some birds. The Quartetto Italiano string quartet. A violin posed in front of the score for the cavatina from Beethoven's String Quartet Opus 130.

Now listen. First, to the second movement of Bach's second Brandenberg Concerto. And then a Javanese Court Gamelan piece, "Kinds of Flowers." And then a Senegalese percussion piece, followed by a Pygmy girls' initiation song, an Australian Horn and Totem song, and "El Cascabel," by Lorenzo Barcelata. Then Chuck Berry's "Johnny B. Goode," a New Guinea Men's House song, "Depicting the Cranes in Their Nest," and then Bach's Partita No. 3 for Violin, followed by a selection from Mozart's *Magic Flute*, "Queen of the Night," Aria No. 14. And then Chakrulo, Peruvian pan pipes, melancholy blues, two flutes from Azerbaijan, and the conclusion from Stravinsky's *Rite of Spring*.

Now, Bach's Prelude and Fugue No. 1 in C Major from Book II of the *48*, and the first movement of Beethoven's Fifth Symphony, followed by a Bulgarian shepherdess song, "Izlel Delyo Hajdutin," and a Navajo Indian Night Chant. Following this, the Fairie Round from Pavans, Galliards, Almains, a Melanesian pan pipe song, and a Peruvian women's wedding song. And then the Chinese Ch'in piece "Flowing Streams," the Indian Raga, "Jaat Kahan Ho," and then "Dark Was the Night." And finally (again) the cavatina from Beethoven's String Quartet Opus 130.

But we're not yet finished. There's more to hear: some whales. A volcano, and mud pots. Falling rain and the sound of surf. Some crickets, some frogs, some birds, and a hyena. An elephant, and a chimpanzee. A wild dog. Footsteps, heartbeats, laughter, and fire. Some tools. And some domestic dogs. The sound of herding sheep. A blacksmith shop. Sawing, riveting, a tractor, and a kiss.

Morse code, a truck, a baby, and the changing of auto gears. Ships. The signs of life signs from an electroencephalogram and electrocardiogram. A horse and cart, and a jet. The lift-off of Saturn 5. A pulsar and a train whistle and a rocket.

And then, salutations from the president of the United States, the secretary of the United Nations, and a whale.

And just in case you're not feeling sufficiently interpellated yet, add greetings in sixty languages, including Sumerian, Akkadian, Hittite, Hebrew, Aramaic, English, Portuguese, Cantonese, Russian, Thai, Arabic, Romanian, French, Burmese, Spanish, Indonesian, Kechua, Dutch, German, Bengali, Urdu, Hindi, Turkish, Vietnamese, Welsh, Sinhalese, Italian, Greek, Nguni, Latin, Sotho, Japanese, Wu, Punjabi, Korean, Armenian, Polish, Netali, Mandarin, Gujoratilla (Zambia), Nyanja, Swedish, Kannada, Ukrainian, Telugu, Persian, Oriya, Serbian, Hungarian, Luganada, Czech, Amoy, Rajasthani, and Marathi.

This is us. This the human greeting to space. This time capsule has one task only: it is to commend on the positivity of humanity. As one wades through the contents of the message, it is a very strange sensation to consider that this is or was intended to be a metonymic distillation of "us." A Rosetta-esque cultural composite.

Apart from the general mendacity of the message itself (the cavatina notwithstanding) it is also with a certain irony that this space-born monument to humanity was conceived as though it were somehow the first message to issue from the Earth. The irony consists precisely in the "noise" that has been ceaselessly beamed outward from the Earth since the time that Guglielmo Marconi started bouncing radio signals around his father's estate. If any message is asserting its importance, it must surely be the redundancy of the barrage of popular programming beaming outwards from the Earth. Laurie Anderson captured this beautifully with the image of dozens of *I Love Lucy* episodes racing outward from the solar system ("Loooooseeee!"). This points to an important parallel between Voyager and the WIPP marker. Just as the Voyager plaque is disingenuous with respect to everything that precedes it, and just as Voyager pretends to operate as though it were *the* disembodied Rosetta-thought of or for humanity, so the marker attempts to convey its message apart from everything that precedes it, and as though it can be a millennial thought-without-a-thinker. A disembodied thought that pertains to a state of affairs.[5]

Here is another one. This is us, too, only in this case we are also the recipient. A number of phonograph records, including two recordings of bird songs, the history of mines, and Artie Shaw. One transcription of the "Premier of Canada." A container of beer, about one quart. A plastic bird, a plastic ash tray, and a beetle plastic ornament and bowl. One vanity makeup mirror with light, a plastic savings bank, and one plastic display case for a watch. A train set, and a cigarette holder. A model air conditioner apparatus, a box of eight plastic samples, and a set of hand scales. One Ingraham pocket watch, a Regen's cigarette lighter, and a woman's Ingraham wrist watch. A sample of gold mesh. A Gen-A-Lite flashlight, an electric Toastolator, and a Monroe calculator. A set of Lincoln Logs. Male and female manikins in glass cases. A desk type telephone instrument dial phone, and ten samples of textile upholstery. Four samples of plated plastics, and a three-cell flash light. Audio scriptions of Thornwell Jacobs's voice. A pencil painting, and a set of Helios game board and pieces. Two carved glass panels, and a set of Bridgeomatic. Two microfilm readers and two microfilms *(Oglethorpe Book of Georgia Verse)*. An obstetrical model, a set of sealed graduates, and a Micarta gear. A package containing six miniature panties, five miniature shirts, three drawers, and a sample plastic radio case. Two Lennox china vases, one blue china bowl, an Emerson radio, and a sample of aluminum foil. Some Technicolor film on display card, an abrasive wheel made of Aloxite, four skeins of rayon, and an electric iron. Two electric lighting fixtures, two acetate shades, and a set of binoculars in leather case. A recording of King Gustav of Sweden, and six transcriptions of the "We, the People" radio show. A small Kodak camera, a plastic drinking glass holder, a sample of catlinite, Lucite, a Schick electric razor, a Comptometer, and a Butterick dress pattern. A Duprene glove, and a silver set. A copy of the *New York Herald-Tribune*, and a Masonic deposit (five badges, and a metal plaque in case, sealed). A glass jar containing two pen holders, three pencils, a slide rule and instructions, a set of colored crayons, a plastic ruler, a fountain pen and pencil set, and six corks. A glass refrigerator dish and cover. A Mazda lamp exhibit, a model Edison's original and a Mazda lamp. A package of assorted wearing apparel, laces and ribbons, stockings, a towel and three washcloths. Framed reproduction paintings of roses and a house. A package of six wood and plastic pictures, and a raffia covered glass powder jar. Soap, assorted hair pins, and some costume jewelry. A glass jar containing a hair bow, a gem razor, a package of blades, a shaving brush, powder puffs, compacts, powder, an eyebrow brush, lipsticks, a hair remover, a toothbrush, rouge, a nail brush, an ivory stick, a pair of manicure scissors, an eyelash curler, five hair curlers, dental floss, tweezers, a package of Mallene, corn pads, eye cup, artificial finger nails, artificial eyelashes, playing cards, and bridge tally cards. A package containing combs, a change carrier, paper cleaning pads, an identification book, dark glasses, a lady's comb, shoe laces, shoulder straps, a flashlight, dice, a cigar holder, and a cigarette holder. Five spools of silk thread, a crochet hook, thimble, needles, rickrack, and bias binding. A sample of oil cloth, a lady's breast form, cellophane dish covers, belts, carbon copy of teletype news, a yellow china bowl, whatnot ornaments, picture hooks, curtain rings and ends, a napkin and napkin ring, wooden forks and spoons set, toy paints, tea bowl, fish hooks, drapery pins, a June bug spinner, curtain rings, a fly, toy watches, pocket knife, smoking pipes, and a bottle of Vaseline. A porcelain figure, two small glass ornaments, a glass coal scuttle, a small glass vase, a glass teakettle, a package of paper clips, cellophane ribbon, measuring spoons, a doughnut cutter, a plastic salt and pepper shaker set, plastic picture frame, and curtain holdbacks. A toy whistle, a golf ball, and a cake of soap. A cover for a milk bottle, a plastic knife, fork, and spoon, a salad fork and spoon, a funnel, barometer, glass container and cover, scouring pad, marbles, outlets, socket plug, switch, pull chain socket, house numbers, rule, can opener, carving knife and fork, screwdriver, grapefruit corer, potato masher, ladle, spoon, pancake turner, asbestos mat, red china plate, girl's head glass bookend, toy automobile, toy stagecoach, and an image of Buddha incense burner. And so on through Donald Duck and a Negro doll. In addition, a lot of microfilmed books, about 800 or so.

All of these things are still here. They're in Atlanta, Georgia, at Oglethorpe University in the Crypt of Civilization (just "down the hall from the University Bookstore, and next to a coffee shop").[6] This, perhaps the most extravagant intentional time capsule, was the work of Thornwell Jacobs and was sealed up and filled with nitrogen on May 28, 1940.[7] The crypt is to remain sealed until AD 8113 (a date Jacobs chose because it was as far in the future as the first recorded date was in the past—which in 1936 was the establishment of the Egyptian calendar in 4241 BC). The crypt was vast (20 by 10 by 10 feet) and well constructed (its walls were lined with porcelain, its floors were two feet thick, and its roof of seven-foot-thick stone).

Just two years earlier, at the moment of the autumnal equinox on September 23, 1938, Westinghouse Electric and Manufacturing Company sank its own time capsule into the fairgrounds at Flushing Meadows, New York.[8] This, the first time capsule to be called a "time capsule," received a tremendous amount of press at the time. It was part of the 1939–40 New York World's Fair, the theme of which was Building the World of Tomorrow.[9] The time capsule contents, though much constrained by the size of the capsule (about the size and shape of a torpedo, oddly), were otherwise thematically similar to that of the crypt. The brainchild of public relations expert G. Edward Pendray, this time capsule engaged with a number of the same problems that the Department of Energy (DOE) would address fifty years later. That is, how to increase the likelihood of the persistence of knowledge of the time capsule, how to locate the site in time and space, and how to account for shifts in language. Keeping in mind that this was basically a publicity stunt aimed at recouping a profit share lost to General Electric, it's all quite extraordinary.[10] The entire project was documented in a book (*The Book of Record of the Time Capsule of Cupaloy Deemed Capable of Resisting the Effects of Time for Five Thousand Years; Preserving an Account of Universal Achievements, Embedded in the Grounds of the New York World's Fair, 1939*), some 500 copies of which were distributed worldwide to public and private libraries, archives, and monasteries.[11] The book gives detailed information about the construction of the capsule; encourages future readers to retranslate the book from time to time as necessary; instructions for the manner in which it ought to be opened; the methods of reckoning the opening date based upon Christian, Jewish, Chinese, Mohammedan, Buddhist, and Shinto calendars; astronomical, heliocentric, ecliptic, and geodetic coordinates; a chapter detailing the method of constructing a metal detector (should the other information fail to properly locate Flushing Meadows); an extraordinary and far-fetched chapter by John Harrington of the Smithsonian setting out a phonetic guide to the English language featuring a mouth map for pronunciation, and a guide to "high-frequency" English vocabulary *(plice dug knock dig hicr)*; a chapter on the methods of contemporary latitude and longitude determinations; as well as letters to the future from the men of "high reputation," Robert Millikan, Thomas Mann, and Albert Einstein.[12]

Of the many operational and conceptual problems with time capsules, there are a few that bear directly on the problem at hand. First, they are meant to be opened at some point in the future. We like sending things to the future, whether these are newspapers stuffed in corner stones or tubes filled with microfilmed books. Notwithstanding this cultural impulse, time capsules generally get lost. We can't find them even if we know they exist, and more often, we don't know they exist and thus never go looking for them. Less obvious perhaps is that the approach to these time capsules is as though cultural language games are or can be entirely divorced from practice, as though language games themselves were not always a question of practice, an abridgement of concrete and specific social practices. There is always a kind of tacit assumption that a sign can be made such that it contains instructions for its own interpretation—a film showing how to use a film projector, a map of the mouth to demonstrate pronunciation, recorded instructions for how to assemble and use a stylus and turntable. Certainly the stakes are fairly low for these endeavors—it hardly matters if they are found and opened on the appointed date, or ever. The problem in the desert, however—a time capsule that must contain the instructions for *its* use, which is to say that it is void of use, that must remain underground for *at least* the legislated duration, that must be remembered correctly, or at least be available to understanding—this problem is different. One way or another, this one must leak.

The Desert

The WIPP project design assumes that the desert is a good place—a particularly good place—to keep a secret. There will be a burial there, and there will be a gravestone, lest we forget. The cadaver in this case will take a great deal of time to decompose. Millennia; too long to comprehend, really. At least too long in the sense that once a duration becomes of a certain magnitude it becomes more or less analogous with a pure future: forever.

The facility is not far from Alamogordo, New Mexico, near the site of the first nuclear detonation of July 16, 1945 (Figure 5). Trinity, it was called.[13] The town of Carlsbad itself is not the site of a famous atomic detonation—although it was very close to the site of the first Project Plowshares detonation in 1961 (more on this later). Until quite recently, Carlsbad's claim to national fame is its proximity

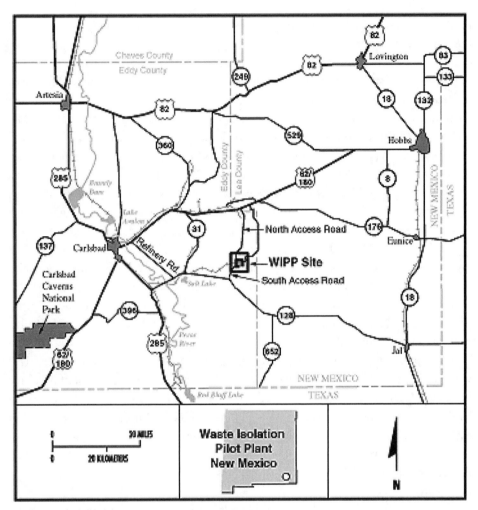

Figure 5. Map showing location of the Waste Isolation Pilot Plant. Map by Gerald Cleal.

to some well-known salt caverns. However, with the development of the WIPP, Carlsbad is now officially on the nuclear map.

The WIPP will permanently store somewhere in the neighborhood of 900,000 specially designed drums of plutonium-based, defense-related nuclear waste: transuranic waste, TRU, as it is known.[14] Waste in the transuranic category may be liquid or solid but generally consists of contaminated protective clothing, tools, glassware, and equipment; in a word, dross.[15] Contrary to the perceived purpose of this site, the WIPP will not significantly reduce the existing U.S. stockpile of radioactive wastes left over from five decades of nuclear weapons research and production at sites across the country. Most of the waste destined to be interred there has yet to be produced.[16]

The WIPP site sits on a 16-square-mile tract of federal land in the exceedingly arid rangelands of southeastern New Mexico. Fewer than thirty people live within ten miles of it. Approximately 50,000 people live in Eddy County, about half of them in the town of Carlsbad, twenty-six miles west of the WIPP.[17] The above ground complex of buildings at the WIPP site is organized around a waste handling building, where the waste containers are unloaded, inventoried, inspected, and prepared for underground disposal. The complex includes a health physics laboratory, an exhaust filter building, emergency electric generators, various office buildings as well as its own fire department, ambulance service, and mine rescue capability. Four vertical shafts allow access and ventilation to the underground portion of the WIPP.

The transuranic waste disposal process that ends in this underground repository is regulated and overseen by a number of government agencies and regulated under a number of laws (the Environmental Protection Agency's [EPA] standards for radiation safety and environmental protection; Department of Transportation and NRC standards for transportation safety; Mine Safety and Health Administration standards for mine safety). In addition, the State of New Mexico Environment Department (NMED) regulates the hazardous chemicals in the WIPP waste and oversees technical aspects of the WIPP through an independent, publicly mandated Environmental Evaluation Group. All work at the site itself is undertaken by the DOE's primary contractor, Westinghouse TRU Solutions LLC (renamed in early 2003, Washington TRU Solutions LLC). Transuranic waste shipments are also subject to certain requirements in the state and local jurisdictions they pass through.[18] The result is a morass of jurisdictional overlaps, and over the past few years, a morass of lawsuits.[19]

The whole question of salt burial is quite fascinating. The official reason why a salt formation is being used as a disposal medium for defense-related, transuranic waste when deep plutonic rock formations are preferable for high-level wastes (as with Yucca Mountain and the Canadian proposal for deep Precambrian disposal)[20] is that most deposits of salt are found in highly stable geological areas with very little seismic activity. The vast sodium chloride formations near Carlsbad were deposited through the evaporation of the Permian sea (late Paleozoic, some 250 million years ago). The salt formation at the disposal site (known as the Salado formation) begins about 250 meters below the surface and extends down some 600 meters. An assumed feature of salt deposits is that they tend to have minimal water circulation that could facilitate the movement of waste to the surface. Indeed, the very presence of salt in formation demonstrates the absence of flowing water. In addition, salt has a high plasticity, making it both relatively easy to mine and prone to "heal" its own fractures. This plastic quality of salt is conceptually knitted into the design of the site; that is, unlike chambers excavated within rock, the salt formations will over time encase the mined areas (and containment rooms) and, ideally, seal radioactive waste from the environment. The persistence of this formation is given to be strong evidence that geological and hydrological activity (earthquakes, subsurface water flow) are a minimal risk to the integrity of the site over the period of time being considered.[21]

The DOE sums it up as follows:

> The proven stability over such a long time span offers the predictability that the salt will remain stable for a comparatively short quarter million years. That's about how long the WIPP-bound waste will take to lose most of its harmful radioactivity and no longer be a threat to the environment. At the depth of the WIPP repository, the salt will slowly encapsulate the buried waste in the stable rock. Relatively small amounts of brine, salt-saturated water, were trapped in the formations millions of years ago. Moisture and salt molecules in the brine will help the recrystallization process to naturally encapsulate the waste in the salt. Meanwhile, salt rock also provides shielding from radioactivity similar to that of concrete.... Stable salt formations offer an excellent repository medium.[22]

Whereas it may be true that seismic activity at the site is minimal, it may well not be true with respect to hydraulic activity. (Pressurized brine has been a problem since the very early stages of the WIPP.) But salt was of initial interest because it was cheap to acquire abandoned salt mines into which to inject waste. In fact this was the reason why, in 1956, the National Academy of Sciences had recommended salt formations as a suitable medium for permanent disposal of radioactive wastes. (After elimination of one potential salt-mine site in Lyons, Kansas, in 1974 the U.S. Geological Survey chose the site near Carlsbad, New Mexico, for exploration. Congress then authorized the WIPP as a demonstration project in a 1979 law, and actual excavation began in 1982.)

The 1992, WIPP Land Withdrawal Act (PL 102–579) withdrew the land from general public use and transferred jurisdiction from the Interior Department to the DOE. It also required the DOE to conduct certain "test phase" activities at the WIPP to demonstrate compliance with applicable disposal requirements. Subsequently, as a result of pressure from antidumping lobby forces, the on-site testing phase of WIPP was redefined, requiring that all testing be done in a laboratory setting. So, as it turns out, the facility was tested to ensure that it performed within specifications without any actual on-site testing as it was originally specified. This was a decisive and catastrophic turn for the antidumping contingent.

Human Interference—Capsulism

This panel member therefore recommends that the markers and the structures associated with them be conceived along truly gargantuan lines. To put their size into perspective, a simple berm, say 35 meters wide and 15 meters high, surrounding the proposed land-withdrawal boundary, would involve the excavation, transport, and placement of around 12 million m³ of earth. What is proposed of course, is on a much grander scale than that. By contrast, in the construction of the Panama Canal, 72.6 million m³ were excavated and the Great Pyramid occupies 2.4 million m³. In short, to ensure probability of success, the WIPP marker undertaking will have to be one of the greatest public works ventures in history.

Frederick Newmeyer, Team A Member

The problems posed by the challenge of permanent disposal were reduced into two more or less discrete problem areas. The first had to do with the material context of the entire site and, more specifically, for the materials that would be used for construction. The question was how the material could be placed into a site so that the likelihood of leakage and migration was as low as reasonably achievable, that the likelihood of future resource extraction (by foreseeable means) was as low as possible, and that the site's potential utility from the point of view of anticipated land use (e.g., less than 12 inches of annual rainfall) was also considered to be as low as possible.

The second problem concerned the temporal security of the site. To this end a task force was established in 1980 by the Office of Nuclear Waste Isolation. The role of the "Human Interference Task Force" was to determine approaches to reduce the likelihood of inadvertent human intrusion into waste repository sites. With the time frame set by the EPA the problem was how to convey the intended message (i.e., go away, danger below) to whoever might visit the site for the period of regulatory concern: 300 generations.

To begin with, the approach was not all that different from that of a time capsule. That is, a free-standing, self-sufficient, meaning-generating (including instructions for the interpretation of its meaning) device; essentially an archive dressed up as a time machine addressed to the year 12000 or so. It is to function as a hybrid device—part time capsule, part memory theatre—which is to say that the sign is an elaborate mnemonic device constructed from highly symbolic information, intended to convey useful information to the future.

In 1980, several scholars were asked to prepare reports on aspects of high-level waste burial and marking systems. These reports established the organizing themes that were to direct and shape subsequent deliberations on both the questions of disposal and marking schemes. The group, known as the Human Interference Task Force, was given a set of working assumptions that included understandings such as there is an ethical responsibility to reduce risk for future persons; the obligation to the future can be discharged if sufficient knowledge is made available to them; future persons are assumed to be capable of breaching any repository design (hence the emphasis on inadvertent intrusion); the focus of design is on the first 10,000 years; and although language cannot be assumed to remain static, future persons can be presumed to have some basic knowledge of physics.[23]

Should future persons elect to breach the repository, they and not the present would be responsible. Therefore, the marker's ethical function works in two directions. It will allow those who should know better to avert the danger. (Although hardly an ethical accomplishment on the part of the present.) And for those who either cannot figure it out or do not care, the present cannot be held ethically negligent. In either case the obligation of the present has been met. The obligation was to be carried out by reducing the likelihood of human intrusion ("interference" as it was then called) and reducing the consequences should an inadvertent intrusion take place. In the first instance—likelihood—the considerations have to do with design and communication features that make intrusion difficult. The reduction of consequences was seen as a function of designing barriers, interring wastes in small "packages," utilizing natural barrier features, and so on.

Some of the works carried out on behalf of the DOE were indeed remarkable, fascinating, and far-reaching. I will mention four texts by way of showing the scope of attempts to define and engage the complexity of the problem.

The most astonishing contribution to the development of the marker was authored by Thomas Sebeok. His work (the only significant contribution from the American semiotic world, and the only contribution that was subsequently reprinted as a popular academic article) was partly a semiotic primer and introduced as well some concepts of information theory—particularly reinforcing the idea of "redundancy" as the key hedge against temporal semiotic decomposition. But for Sebeok the problem was bigger than this. Foremost he viewed it as a problem of the sign's stability through time, the felicity of the sign. His first recommendation was

> that information be launched and artificially passed on into the short-term and long-term future with the supplementary aid of folkloristic devices, in particular a combination of an artificially created and nurtured ritual-and-legend. The most positive aspect of such a procedure is that it need not be geographically localized, or tied to any one language-and-culture.[24]

The idea was that the present would design a kind of epistemological false trail such that people would be disinclined even to visit the site. And this disinclination would not necessitate any knowledge of the meaning of the site or of the nature of the materials interred. "A ritual annually renewed can be foreseen," he said, "with the legend retold year-by-year (with, presumably, slight variations)."[25]

However, the manufacturing of a *new* tradition designed to secure the site, fabricating and hacking a contemporary mythological deep structure, was, in Sebeok's view, insufficient. In addition, he saw the need for a transhistorical assembly of experts. The "truth" of the site

> would be entrusted to—what we might call for dramatic emphasis—an "atomic priesthood," that is, a commission of knowledgeable physicists, experts in radiation sickness, anthropologists, linguists, psychologists, semioticians, and whatever additional expertise may be called for now, and in the future. Membership in the "priesthood" would be self-selective over time.[26]

Thus constructed, the nuclear Templars would be charged with mythological supervision and the production of metamessages as necessary. Should future generations fail to obey the imperative to care for the site,

> the atomic priesthood would be charged with the added responsibility of seeing to it that our behest, as embodied in the cumulative sequence of metamessages, is to be heeded—if not for legal reasons, then for moral reasons, with *perhaps the veiled threat* that to ignore the mandate would be tantamount to inviting some sort of supernatural retribution.[27]

As profoundly cynical as Sebeok's proposal may appear, he nonetheless saw very clearly the futility of merely launching a sign into the future. The pragmatics of the sign—the work that it must do—at least makes Sebeok's proposal a plausible thought experiment.

Some of the research was concerned less with the sign itself than with cultural supports to maintain knowledge of nuclear wastes and their locations. Weitzberg wrote a paper entitled "Building on Existing Institutions to Perpetuate Knowledge of Waste Repositories." His work focused on techniques for deploying existing systems of "information" archiving (libraries, online databases, National Archives, maps, geodetic surveys).[28] The deployment of existing practices of knowledge, from maps to periodic tables, has since become an important feature of the marker design proposals. There are two things of note here. First is the idea of an archive as a place where knowledge can survive independent of the culture that produced it. And second is the implicit assumption that information (i.e., data) is equivalent to knowledge to begin with. Knowledge, in other words, without a knowing subject.[29] Nonetheless this addressed an important, if obvious, concern that the marker itself could not be solely counted on as the source of information about the wastes.

From the concern for the sign and question of the archive, another focus of research was about the human. The work of Percy Tannenbaum focused on what he saw as universal characteristics of the human perceptual makeup.[30] Determining these basic elements of human perception, whether facial expression or fear reactions to menacing figures, became a prominent theme in discussions concerning the philosophy of the marker design; equal parts Jungian and behaviorist, a design that can be propped up by an essential human dimension became a seductive proposition. In other words, the question of a sign that could iconically point to human fear (more on this later).[31] This line of inquiry attempts to outflank or preempt the semiotic and archival questions by appealing to human perception at a "deep structure" level.

The last area of research I will mention amounts to a kind of pragmatic, historically based semiotics. The archaeologist Maureen Kaplan did some very significant work in this area in that it contextualized the problem of a marker *as* a historical problem (i.e., the transmission of meaning across time as itself an historical question).[32] In order to address this, she suggested a four-level taxonomy for how information should be conveyed, from the very simple to the very complex: something is here; it's dangerous; it's dangerous and here's why you should go away; and, here's some detailed symbolic information. This taxonomy was not only about message redundancy, but about an assumption of changing interpretive strategies, languages, and symbolic competencies.

A pragmatic taxonomy of layered messages, with the theoretical support derived from Sebeok's flagging of redundancy, wrapped up in the idea that there are culturally neutral, transhistorical signs that would invoke human fear, etc., informed almost all subsequent design ideas. But it is important to see that at this point in the development of the marker project, the problem was granted a considerable complexity. As government funded research, this was far from standard fare.

Futures Panel

Following on the work of the Human Interference Task Force, the DOE began a process through which they would "define the criteria which will be used to decide what kind of passive markers can be used at the WIPP to significantly mitigate the effects of the human intrusion scenarios on performance assessment."[33] A new interdisciplinary working group—Futures Panel as it was called—was assembled in 1990 and given the rather unbelievable task of identifying the range and possible configurations of future societies that might occur in the region of the WIPP within the next 10,000 years, in order to establish the modes and probabilities of various inadvertent intrusion scenarios.[34]

> Because the regulatory period for the WIPP spans 10,000 years, societies different from our own may encounter the buried radioactive waste left by us. Even though the potential risk associated with radioactive waste decreases with time, it is still necessary to consider possible future societies when designing markers and obstacles to prevent human intrusion. One approach is to create alternative futures for the development of society. These alternative futures can be constructed by considering alternative projections of basic trends in society. These trends may include population growth, technological development, and the utilization and scarcity of resources, among other factors. Overwhelming these factors in the possible impact of human intrusion are events that modify, or reinforce the development of society. Such events may include nuclear war, disease, pestilence, fortuitous discovery of new technologies, climatic changes, and so forth. The creation of a reasonable set of alternative futures provides the first step in evaluating the types and likelihoods of intrusive activities. It is not possible, however, to ensure that all possible futures are considered. It is not even reasonable to assume that humans can conceive of all possible future societies. The farther into the future we delve, the less complete these alternative futures are likely to be.[35]

The Futures Panel produced an unbelievable set of probabilistically based future society and intrusion scenarios, keyed to 100, 1000, and 10,000 years in the future. The questions of cultural (dis)continuity and technological change played a critical role in the panel's deliberations, but so too did questions of geo-political and linguistic shifts, radical fluxes in population distribution and density, changes in rates of literacy, global catastrophe (from thermonuclear war to extra-terrestrial attack), questions of cultural memory, and changes in regional and global rates and types of resource utilization. Some imagined scenarios were quite fantastic:

> A Feminist World, 2091. Summary: Women dominated in society, numeri-cally through the choice of having girl babies and socially. Extreme feminist values and perspectives also dominated. Twentieth-century science was dis-credited as misguided male aggressive epistemological arrogance. The Femi-nist Alternative Potash Corporation began mining in the WIPP site. Al-though the miners saw the markers, they dismissed the warnings as another example of inferior, inadequate, and muddled masculine thinking. They pene-trated a storage area, releasing radionuclides.[36]

Other scenarios posed questions such as what might happen if there developed a quasi-religious, radically relativist cult based upon Kuhn's *Structure of Scientific Revolutions* and Marcuse's *One-Dimensional Man*.[37] For the Futures Panel this possible world scenario was one where the "Markuhnians" (or the Markuhnian Conspiracy) attempted to lead an antitechnoscientific cultural revolution. In an at-tempt to locate sacred buried scrolls, the Markuhnians—dismissing the warning monuments at the WIPP site as the arbitrary production of an incommensurable version of reality—breached the repository and released a geyser of radioactive brine.

In addition to thinking about technology and various cultural discontinuities, some very interesting observations about memory were made. In particular, the idea that without memory the meaning and importance of any mark-ing systems used at the WIPP site were (in principle) impaired. By way of example the report mentions the Project Gnome detonation of December 10, 1961.[38] Gnome was part of the Atoms for Peace project established in 1957 called Project Plowshares (i.e., "and they shall beat their swords into plowshares," Isaiah 2.4). The stated ob-jectives of the project—which involved 27 detonations between 1961 and 1973— were to discover peaceful uses for the bomb; for example, harbor creation, resource extraction, earth-moving, heat and isotope production. It was the last two applications that were of primary interest in the 3.1 kiloton Project Gnome detonation. Gnome was the first explosion in the program and received a great deal of media attention at the time. It took place about 25 miles southeast of Carlsbad—about 6 miles from the WIPP. Gnome, designed as an underground (1216 feet) and fully contained detonation, was a catastrophe. The tunnel in which the explosive was detonated was supposed to be self-sealing—instead the explosion immediately vented to the atmosphere. Today, attached to a small cairn, there are two plaques; one, a somewhat corroded plaque describing the project (though not mentioning the unintended result), and another prohibiting excavation and drilling, for an unspecified period, "at any depth between the surface and 1500 feet."[39] The Futures Panel suggests several strategies that—while one may not agree with the specifics—show how social practices would need to be built up around the site. For example, as a kind of nod to Sebeok's Templars, they suggest setting up graduate student fellowships that would require site visits in perpetuity and creating a twenty-five-year review process where intrusion scenarios are repeatedly reevaluated.[40]

From the Futures Panel one can see a glimmer of a critique with respect to an overreliance on markers: they can be lost; they may become incomprehensible; they can't teach risk analysis; they do not assure a continued governmental stewardship (indeed they may facilitate the opposite); they do not prevent a loss of memory about the site and its purpose; and they do not necessarily deter land development (industrial or otherwise) that might threaten the site.

Markers Panel

The next development was the recruitment in 1990 of the Markers Panel.[41] The disciplinary areas of expertise represented were materials science, architecture/environmental design, anthropology, linguistics, archaeology, astronomy, communications, geomorphology, scientific illustration, semiotics, and environmental engineering. Here again the problem was approached as a kind of time capsule puzzle. They were interested to a certain extent in the materials that would be utilized and their likely durability and so on, but their principal concerns were how to design a system of marking that would convey the danger of the site to the future.

It was clear to the teams that to rely upon language—written texts—to carry the burden of meaning was dangerous. But both teams also felt that textual accounts of the area were necessary at least in the near future (100–500 years). The presumption of linguistic mutation and perhaps even the emergence of unique languages presented a difficult problem. Both teams more or less followed the leveled message taxonomy mentioned above, and both acknowledged the idea that linguistic indeterminacy did not foreclose the use of signs. However, apart from the informational aspects of the design, both teams approached the problem of the marker as though the site itself could be made to *look* dangerous. The design wouldn't in fact be dangerous; it would signify danger—that is, the phenomenology of the place would somehow equal danger, and the site's contents would be both qualitatively and quantitatively explanatory.

The first team founded their design upon the conviction that "communication technology cannot bypass the problem of the certain transformation and succession of cultures, but the use of fundamental and enduring psychology can"; and "the entire site must be experienced as an integrated system of mutually reinforcing messages, and designed accordingly."[42] Thus, the object of their design work was to apprehend the fundamental and enduring and to deploy these sign elements in an integrated fashion.

> Modern understanding of the communications enterprise shows that there can be little separation of the content of a message from its form, and from its transportation vehicle. They affect each other, and all of it is message. McLuhan and Fiore take that even further, arguing that "the medium is the message." Given this, rather than our attempting to first articulate messages, then to select their form, and then to design their vehicle, we choose to do as much of this simultaneously as is reasonable, attempting to accomplish:
>
> > —a *Gestalt*, in which more is received than sent,[43]
> > —a *Systems Approach*, where the various elements of the communications system are linked to each other, act as indexes to each other, are co-presented and reciprocally reinforcing, and
> > —*Redundancy*, where some elements of the system can be degraded or lost without substantial damage to the system's capacity to communicate.
>
> > Everything on the site is conceived of as part of the message communication... from the very size of the whole site-marking down to the design of protected, inscribed reading walls and the shapes of materials and their joints. In this report, the various *levels* of message content are described, as is the *content* of each level, the various *modes* of message delivery, and the most appropriate *physical form* of each.[44]

Accordingly, the design they developed is roughly as follows. The Level I message would be the site itself; the site as a gestalt of danger. The organization of the elements and the phenomenology of the place for its witness—any witness—would be:

> This place is a message... and part of a system of messages... pay attention to it!
>
> > Sending this message was important to us. We consider ourselves to be a powerful culture.

This place is not a place of honor...no highly esteemed deed is com-memorated here...Nothing valued is here.

What is here was dangerous and repulsive to us. This message is a warning about danger.

The danger is in a particular location...it increases towards the cen-ter...the center of danger is here.

The danger is still present, in your time as it was in ours.

The danger is to the body, and it can kill.

The form of danger is an emanation of energy.

The danger is unleashed only if you substantially disturb this place physically. The place is best shunned and left uninhabited.

The Level II message would be inscribed on surfaces throughout the marker area in the six languages of the United Nations and "a local language such as Navajo." It would read as in Figure 6.

Pictured to the left of the text is a likeness of the head from Edvard Munch's *The Scream*, and to the right, a face picturing "nausea" from Eibl-Eibesfeldt's *Human Ethology*.[45] The work of ethologist Iranäus Eibl-Eibesfeldt moved smoothly (as exemplar and confirmation, simultaneously) into this structure of uni-versal meaning—faces expressing horror, revulsion, fear, pain, anguish providing sup-port for an essential, phenomenological, or haptic mode of perception.[46]

The figure of the monument is conceptualized in a very interest-ing manner by those charged to think about it. They have misunderstood the mon-ument as something that conveys meaning. And they have been able to reach this point in their thinking through the assumption that some things just mean what they mean. Which is of course very different from saying that things simply signify before we know what they signify, that the signified is given without being known. "Your wife looked at you with a funny expression. And this morning the mailman handed you a letter from the IRS and crossed his fingers. Then you stepped in a pile of dog shit....It doesn't matter what it means, it's still signifying."[47] That is, they began with the assumption that certain physical forms have the capacity to convey extralinguistic, stable pancultural meaning. And further, that this capacity is based on an "enduring human propensity to experience common and stable meanings in the physical form of things, including the design of landscapes and built-places."[48] To this claim, they would point for example to the "powerful feelings invoked" in the viewer in the presence of the paintings at Lascaux or Altamira. Accordingly there is an assumption of a capacity on the part of the object (i.e., to be a vehicle of trans-mission), and a propensity on the part of the human (i.e., to perceive and experience

DANGER
POISONOUS RADIOACTIVE ☢ WASTE BURIED HERE
DO NOT DIG OR DRILL HERE BEFORE 12,000 A.D.

Figure 6. Proposed Level II message to indicate the presence of danger.

objects similarly)—and the result is a notion of archetype. A fortuitous notion since it seems to offer a way to avoid all the messy problems of the instability of meaning, symbolic ambiguity, and temporal semiotic decay.

The Level III message is to give a textual explanation of the site and its purpose, but it contains no detail or specialized language concerning the contents or the mechanisms of threat.

> These standing stones mark an area used to bury radioactive wastes. The area is...by...kilometers (or...miles or about...times the height of an average full-grown male person) and the buried waste is...kilometers down. This place was chosen to put this dangerous material far away from people. The rock and water in this area may not look, feel, or smell unusual but may be poisoned by radioactive wastes. When radioactive matter decays, it gives off invisible energy that can destroy or damage people, animals, and plants.
>
> Do not drill here. Do not dig here. Do not do anything that will change the rocks or water in the area.
>
> Do not destroy this marker. This marking system has been designed to last 10,000 years. If the marker is difficult to read, add new markers in longer-lasting materials in languages that you speak. For more information go to the building further inside. The site was known as the WIPP (Waste Isolation Pilot Plant) site when it was closed in....

The Level IV message is the message with the most content and detail concerning the site. They propose two rather long texts as possible variations on the Level IV message. Both detail the nature of the waste, the mechanisms of toxicity, the depth at which it is buried, maps of the site, a periodic table, star maps to indicate the decline of radioactivity by showing the passage of time, the location of all other known waste sites, a description of the symptoms of radioactive sickness, and instructions to reinscribe the surfaces of the marker with updated information.

Some of the proposed information and particularly its mode of presentation is very difficult. Consider Figures 7 and 8. The first diagram is to assist in the location of the sites of waste throughout the globe. The outer circle is to indicate longitude, and the inner circle, latitude (I still have a difficult time with this one).

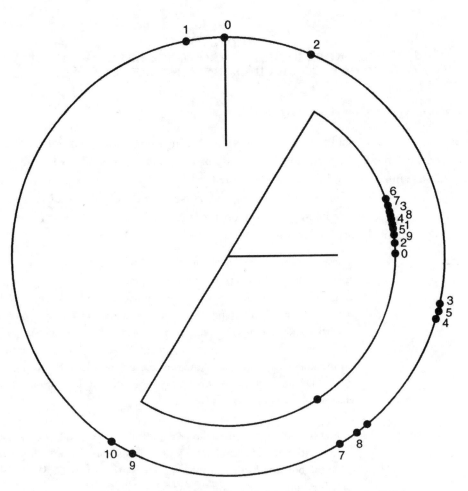

Figure 7. Level IV diagram to show the location of global waste storage facilities.

As with all Level IV messages, there would be text to support this image, but none-theless, it is a startlingly abstract picture of the globe.

I find Figure 8 (a Level III message conveying "basic informa-tion" and thus by definition "easier" than Figure 7) equally perplexing. They write: "To those not able to understand any languages, this diagram [Figure 8] will indicate both the epoch of burial, and the period of danger."[49]

The report of the second team also used the idea of multiple levels of messages, though their design was less elaborate. They specified that earthwork berms be constructed around the perimeter of the site to an elevation of thirty feet.

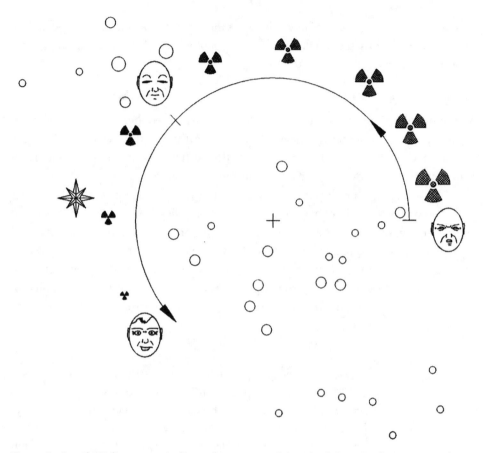

Figure 8. Level III diagram to indicate the passage of time (and thus the decrease in radio-activity) as an indexical function of the "movement" of stars about Polaris. The retrograde movement of the diagram passes from the anxiety face to the happy face, and the (as yet unknown) international symbol for burial diminishes in size.

The overall outline of the earthwork would be either a skull and crossbones or a nuclear trefoil. One-piece granite monoliths (ten by twenty-five feet) would be arranged at intervals around the perimeter. The sheltered surface of each monolith would be used for various Level II inscriptions, and their total number would be a power of two so the original configuration of the ring could be inferred by future investigators. A structure would be placed at the center; this would be the repository for the Level IV information. Small time capsules would be buried around the site containing Level II and III information. The contents of the time capsules would be such things as "durable tablets," samples of wood for carbon-14 dating, and small-scale cross-sectional models of the geological substrate, mine shafts, and depository rooms.

Mr. Yuk

Within the deliberations of the Markers Panel, one can see there were two principal areas of conflict concerning design philosophy. Although the two teams basically agreed to the majority of design concepts, they differed concerning the nature of sign units to be deployed and as to the question of a center.

The dispute over sign units turns on the question of whether and how much to rely upon various types of "graphics." The second team based their design upon the assumption that pictographs have a pancultural character and as such ought to be deployed in order to display a narrative concerning the development of the site and the danger of intrusion. Because they felt that "symbols have more emotional content than other signs," they recommended that the choice of symbols should be left for future researchers, and in any case, should be "defined pictographically."

Writes Peirce:

> A *Sign* [representamen] is anything which is related to a Second thing, its *Object*, in respect to a Quality, in such a way as to bring a Third thing, its *Interpretant*, into relation to the same Object, and that in such a way as to bring a Fourth into relation to that Object in the same form, *ad infinitum*. If the series is broken off, the Sign, in so far, falls short of the perfect significant character. It is not necessary that the Interpretant should actually exist. A being *in futuro* will suffice.[50]

It is difficult to make semiotic sense from all of this. What they attempt to say is that an image such as "Mr. Yuk" (which has apparently been adopted—though from my point of view without much evident success—as the international symbol for use as a warning for children on prescription medicine labels) has less inherent ambiguity than signs such as the international biohazard symbol or the standard nuclear trefoil (Figure 9). In other words, Mr. Yuk—and never mind the "Oriental" resonance of the name or indeed the caricature of the face—is taken to be a less conventional, more motivated, and natural sign. That is, it is both iconic and indexical in a very Peircian manner. What distinguishes an icon is that it operates via similarity. The sign vehicle (the representamen, as Peirce called it) is *like* the object, has a similarity with the object. What we might mean by "is like" is not a

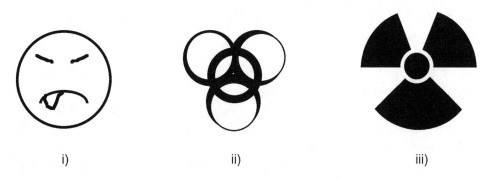

i)   ii)   iii)

Figure 9. Variations on pictographs: "Mr. Yuk" (left) is presumed to have less inherent ambiguity than either the international biohazard symbol (center) or the standard nuclear trefoil (right).

simple affair, and in the end one could claim that everything is "like" everything else, but the point here is that Mr. Yuk stands as a particular sort of sign that, in pointing to its object, generates the interpretant it does because the sign vehicle (Mr. Yuk) and the object really are similar in some significant sense.[51] But Mr. Yuk is also described as though it has an exclamatory quality of pointing to its object: This is bad! It is as though the Mr. Yuk sign was not merely similar to its object but also connected to it—that is, an index. ("A sign which denotes a thing by forcing it upon the attention is called an index."[52]) But even still, Mr. Yuk can only work (have the desired function) when it is also in a place where its object exists. (At the level of the index, it must work this way.) Its associations of similarity, the understanding of its exclamatory quality, and the awareness of the continuity of an indexical function are determined by force of habit (or law) and are consequently in the mode of a (conventional) symbol as well. So, like most signs, Mr. Yuk is complicated. It's not pure, and while it may have some qualities of indexicality and iconicity, these qualities are interpretively mediated by very conventional means of cultural (linguistic) competence.

The other two signs, the biohazard and trefoil, are taken to be weaker signs, they are simply conventional. Although from the point of view of the sign, they are in fact stronger, not weaker. These are more obviously symbols in Peirce's sense: "a sign which owes its significant virtue to a character which can only be realized by the aid of its Interpretant."[53] That is, the understanding of such signs— which for Peirce would be the interpretant—is due only to convention, to habit, to some competence that allows us to recognize that we are meant to make a connection. Nothing about the symbol itself and its relation to its object determines our understanding of it as a sign. It operates not by similarity, nor does it have any actual connection to its object. It is a sign only to the extent that we agree that it is, only to the extent that we see it as such.

The very idea of a pancultural sign—and one might as well say transhistorical as well for the pancultural to make any sense—that is, a sign meaning the same thing to everyone, requires that a sign, any sign, operate primarily in one of two modes. That is, either as index or icon. But as Peirce cautions, a pure icon can "convey no positive or factual information; for it affords no assurance that there is any such thing in nature."[54] While a pure index would have the being of a present experience, it would have to forcibly intrude upon the mind as such, regardless of to whom, or where, or when this encounter took place. "Icons and indices assert nothing."[55]

> If an icon could be interpreted by a sentence, that sentence must be in a "potential mood," that is, it would merely say, "Suppose a figure has three sides," etc. Were an index so interpreted, the mood must be imperative, or exclamatory, as "See there!" or "Look out!"[56]

The wish for the waste to be made secure by the emplacement of signs that, in their very purity, have a sociobiological behavior or compulsion guides us into a look at the larger design ideas.

Two monument schemes proposed by the second team are shown in Figure 10.

The skull and crossbones motif (Figure 10b) was suggested by Carl Sagan in a letter he wrote to Sandia Laboratories to decline his participation in the project. Of it, he wrote:

> I think the only reason for not using the skull and crossbones is that we believe the current political cost of speaking plainly about deadly radioactive waste is worth more than the well being of future generations.[57]

Sagan claimed that the skull and crossbones was *the* pancultural sign to signal grave danger (from the lintels of cannibals, to SS and biker gang insignia and pirates). There is perhaps a bit of cultural blindness here. Particularly in light of the proximity to Mexico (and in general, the Spanish American population in the Southwest), one need only point to the festivities around el Dia de los Muertos. Particularly in urban contexts, and in the Southwest, the skull and skeleton iconography (indeed, the unbelievable array of extraordinary Day of the Dead kitsch) just simply invalidates the whole idea of the skull and crossbones as a deterrent. As an enticement, perhaps, but warning, no.

In any case, the first team identified what they saw as specific philosophical difficulties inherent in the use of what they called "graphics." They identified the danger of ambiguity (citing the thematic apperception test[58]), the danger that graphics would be removed from the site (the ambiguity of art), and in general the culturally restricted manner in which graphics may operate as signs. Not entirely, though. The only "universal" sorts of graphics they recommend for use are that of the human face in various "emotive" or affective states (e.g., "pain, anger, disgust, fear"). Signs, in other words, that convey affect. (It is unclear whether Mr. Yuk would apply here or not.)

Whereas the second team advocated extensive use of cartoon pictographs throughout the marker site (Figures 11 and 12), the first team wrote a strongly worded section warning that written language has a higher probability of being understood. Their argument was simply that the *symbols* associated with, say, alchemical texts are more obscure today than are the alchemical texts themselves:

> We suspect that 500 years from now, it will be correspondingly easier to uncover the meanings of the English words "radioactivity" and "hazardous waste" than of the symbols now used to denote them.[59]

What this amounts to saying is that some symbols, or replicas, are more conventional than others, which is most certainly true. What this does not say, however, is for whom the meaning means.

In Figure 11, the second team shows how what they call symbols can be used algorithmically to form equivalencies. That is, they are to be used to predict a future state. Figure 12 shows the application of the same idea to the definition of a single symbol. It is, I think, quite clear how these are not the same procedures. In the first case, it must be read as a series of transpositions based on equivalence (trefoil → atom, poison → Mr. Yuk, prohibited cross → conventional crossed-out X). Nothing takes place in Figure 11—it is an attempt at the derivation of rules. Whereas in the second case (Figure 12), the procedure is not one of assigning equivalence at all. Rather it must be *read*, as a narrative, top to bottom, as a temporal sequence in which there is action and something takes place. But what exactly takes place is a tough call. For example, either there is linear perspective involved, or the tree grows considerably larger. Rather than an innovation in the technique of picturing, this could be a technique for showing the passage of time. In addition, one might wonder what becomes of the small monument that accomplishes the transfer of the graphic onto the T-shirt; is it the modification of the body, a scarification, perhaps? Or a kind of magical silk-screening?

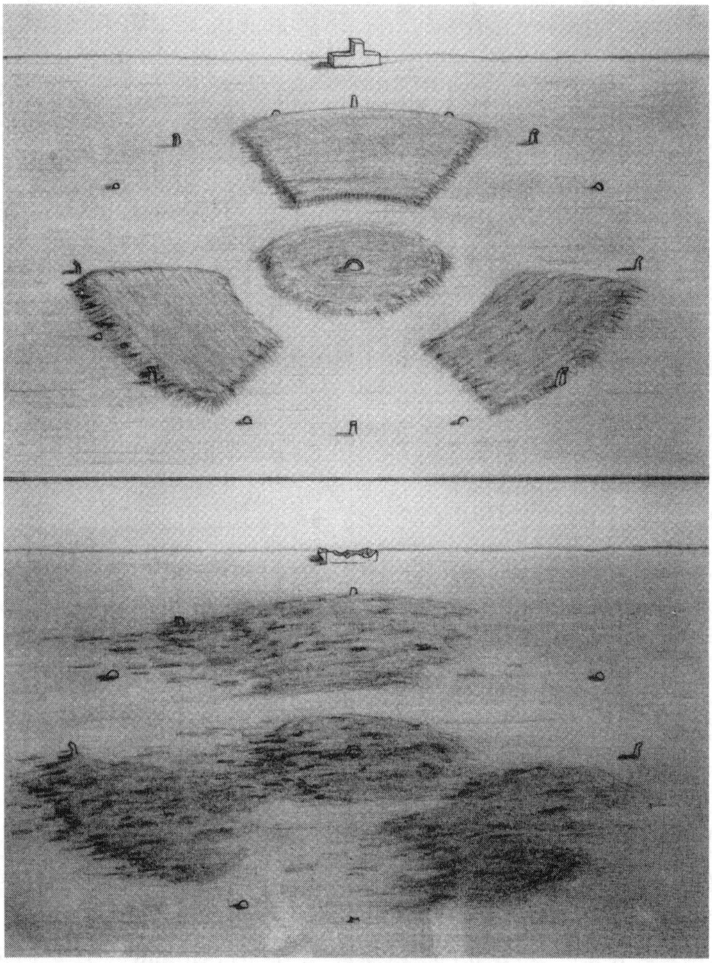

Figure 10. (a) Marker concept proposal showing nuclear trefoil pattern at time of construction (top) and after five thousand years. (b) Marker concept proposal showing skull and crossbones design at time of construction (top) and after five thousand years. Art by Jon Lomberg.

Pass It On

We begin with this: This place is dangerous. Go away!

A simple utterance. A claim about a state of affairs, and a command. Its performative dimensions are contingent upon it being understood and also that this understanding is communicable. In other words, it is not just I who must understand. But also you and you and you.

b

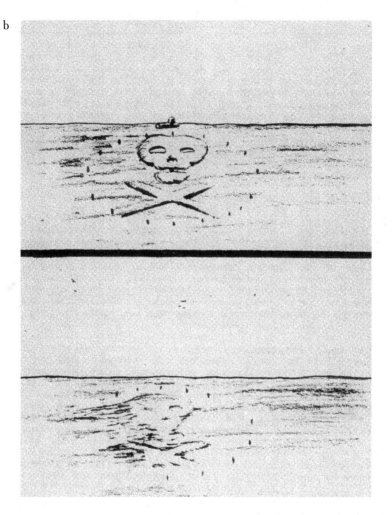

Now, via a little illustrative trick used by Marcos Novak, we take the phrase and do a machine translation into French[60]:

Cet endroit est dangereux. Allez-vous—en!

Voila. This is an unfortunate turn. Now back to English:

This place is dangerous. Go you—in!

Same problem. Now into German:

Dieser Platz ist gefährlich. Gehen Sie—innen!

Figure 11. Proposed series of pictographs to define the equivalence of symbols.

And then back to English again:

> This workstation is dangerous. Go—inside!

Same problem at the level of the command, but the state of affairs has changed. The site that was dangerous has disappeared. It has been transformed into a techno-apparatus—also dangerous—something called a workstation. Next, Italian:

> Questa stazione di lavoro è pericolosa. Andare—all'interno!

Then back to English again:

> This station of job is dangerous. To go—to the inside!

The abstraction increases. Now the workstation has become just a station, making ambiguous what it means. Station in the sense of one's station in life? Portuguese:

> Esta estação do trabalho é perigosa. Para ir—ao interior!

And back again:

> This station of the work is dangerous. To go—to the inward!

Figure 12. Pictograph series to define radiation trefoil symbol.

But now the inside, as place, has been replaced by the direction, inward. From a place to a workstation to a job, and now to work. Now the final translation, and then English:

> Esta estación del trabajo es peligrosa. Para ir—al interior!
> This station of the work is dangerous. In order to go—to the interior!

The claim and the command are starting to look a lot like a koan.

The Center

The second point of dispute between the two teams concerned the center of the site. As mentioned above, the second team incorporated a central structure into their design as a principal focus for the site. They wrote:

> Central placement of [the] rock shelter would draw future visitors through the encircling earthwork and the ring of monoliths to the center of the marker, where inscriptions inside would carry pictographic, linguistic, diagrammatic and scientific information. The designed shape itself would attract people to the structure.[61]

Not far inside this proposal lurks the outline of a disaster theme park, apocalyptic infotainment not far from the scenarios proposed by the Futures Panel as a way of designing against memory loss.[62] The other team suggests a very different design philosophy:

> For human beings, making a center ("here we are") is the first act of marking order (Cosmos) out of undifferentiation (Chaos). The meanings of center have always been of a highly valued place ... the holy of holies; the statue centered within the temple; the dancing ground; the sacred place as the physical and spiritual center of a people. . . . In this project we want to invert this symbolic meaning, to suggest the center is not a place of privilege, or honor, or value, but its opposite. In symbolic terms, we suggest that the largest portion of the Keep, its center, be left *open*, and few (if any) structures placed there, so that symbolically it is: uninhabited, shunned, a void, a hole, a non-place.[63]

This is a very interesting idea; it has a cleverness that seems appealing. And perhaps in some sense it is correct. But one can also see the almost impossible task that this would create. Only by abstracting the site itself from its immediate desert context could this argument about symbolic inversion make sense. In other words, whether the installation itself has a center would seem to be incidental to the fact of the installation standing alone in the desert. Its very presence would

Figure 13. Spikes Bursting through grid marker design. Concept by Michael Brill and art by Safdar Abidi.

assert itself as a prominent center. An ambiguous assertion to be sure, but clearly an assertion of its presence. (The proposed system of representation would allow the WIPP to usurp Greenwich and the Equator to become 0,0, the center par excellence.)

The designs that were most thoroughly considered at the time—in addition to the berm design mentioned above—were all described as unified in the sense that they "utilize archetypal images whose physical forms embody and communicate meaning."[64] But mostly, it seems to me, they are just beautiful. Their names betray the anxiety that underlies their physical ambiguity.

The designers undertook to create designs that looked like danger, mimetic signs that look dangerous. Therefore they gave them dangerous sounding names: Landscape of Thorns (Figure 13), Spike Field, Spikes Bursting through Grid, Leaning Stone Spikes, Menacing Earthworks, Black Hole (Figure 14), Rubble Landscape, Forbidding Blocks.

Like their names, the shapes of these designs are said to "suggest danger to the body . . . wounding forms, like thorns and spikes, even lightning."[65]

Significantly, though, none of these designs really *are* dangerous. They may provide an inhospitable environment for certain activities, it could be

Figure 14. Black hole marker design. Concept by Michael Brill and art by Safdar Abidi.

difficult to get around on them, difficult for machinery perhaps, not a place you might want to spend time. They may indeed provide a real challenge for one who may want to *be* there. But they do not present real danger. The only exception, and the thing that might contain the seed of a really important idea, is the Black Hole design (Figure 14). As they describe it:

> A masonry slab, either of black basalt, or black dyed concrete, is an image of an enormous black hole; an immense nothing; a void; land removed from use with nothing left behind; a useless place.... The blackness absorbs the desert's high sun-heat.... The heat of this slab will generate substantial thermal movement.[66]

If we pass over the description of its "nothingness," for it surely is not nothing, the interesting idea that marks this as a singular moment in design thinking is that it is a sign that will hurt. It doesn't *refer* to pain. In fact, its distinctive features do not point at all. It has a far more intimate connection with bodies—its design is such that it would deliver pain. In a small but provocative way, it short-circuits the need for representation, by fusing itself as a thermal sign. The sign hurts.

The sign and the body work together as a pain-making assemblage. Even in this, though, even in the directness of the connection with bodies, we are left with either an ambiguous index—"*that* hurts, but I don't know why"—or simply a distinctive event—both impressive nonetheless. For even if it is grasped as a sign, if it is understood as not fully realized in the pain it can produce in its actualization with a body, it still says nothing clear about what it then might mean. It says only that it means *something*. In the latter case, which in the end may come to the same, there is an event, an encounter, partially closed in on itself, and partially in the form of a question: Why? The distinctiveness of this design—an awareness that it contains the idea of a radically *different* kind of sign—seems to have passed unnoticed.

As if an afterthought, another kind of sign emerges in the final paragraph of the second team's report. In this instance it is a dissonant and moanful sign. A sign that does something.

> Communication of the basic Level I message could also take place through sound.... The effect of the various sounds generated should be consonant [so to speak] with the overall site design, namely a place of great foreboding. Indeed sounds that can readily be generated by long-lasting aeolian structures turn out often to be dissonant and mournful.[67]

A long and low and plaintive cry in the desert.[68] An Aeolian sign. A sublime image of the deep irony of the wind's indifference.

The To-Do List

No one has any idea where the desert project will end and whether and how the DOE will make good on their spatial and temporal responsibilities.[69] Right now, no one is really thinking much about the long-term security of the site. For the time being the DOE is keeping everyone occupied with hearings having to do with modifications to their operating permit. They've made requests to change procedures for surface handling, testing of wastes, and atmospheric sampling in the waste containers, and they've made requests to change the length of time that materials may be stored prior to being placed underground and for the kind of wastes they may process.

In April 2001, under the new Bush administration, Secretary of Energy Spencer Abraham decreed that the DOE must be run more like a business and simultaneously predicted a rosy future for the National Transuranic Waste Management Program. This requires cutting costs, reducing approval procedures, standardizing all manner of criteria, and ultimately doing more for less. At the WIPP this set off a number of changes to the short-term and long-term operation and administration procedures (adopting corporate organizational structure and current business tools, a new focus on productivity, and so on) and foremost an attempt to reduce the unit cost of handling waste containers and to increase the number of waste shipments processed per week.[70]

The promising, if incredible, fanciful thinking of the Markers Panel and the Futures Panel, stalled right at the point in history when concepts threatened to break free, where something interesting could have happened.

In 1992, Congress required the EPA to ensure the safety of the WIPP site. In response, the EPA built a set of disposal standards and required the DOE to demonstrate that the WIPP would in fact meet these standards. In February 1996, the EPA followed those general standards with more specific Compliance Criteria related to the WIPP site itself. The purpose of the Compliance Criteria document was to clarify the requirements of the radioactive waste disposal regulations and set out the requirements for the DOE's "Compliance Certification Application."

In October 1996, the EPA received the DOE's impossibly huge "Compliance Certification Application" (some 84,000 pages) and immediately began its review for completeness and technical adequacy and, shortly thereafter, public comments and public hearings. The task of the EPA was to decide whether or not the DOE had met with the criteria as previously set out.

The DOE—at least as far as this document was concerned—opted for a lowest common denominator solution of compromise. By making simpler arguments for simpler designs, the DOE attempted to ensure the project would be granted approval.

The passive institutional controls they have proposed and the design concepts that have been advanced to the EPA have come under the numbing constraints of materials design and cost. And the EPA itself has backpedaled on the assurance criteria that the DOE was charged to demonstrate.

Virtually all the other questions that may have been posed have, for now, been excised from the realm of the problem. For instance, there will be no discussion about how the waste might be kept aboveground in isolated and monitored storage facilities, a manner of storage where the knowledge about practices around the waste could be an ongoing matter of technical, social, and ethical concern. Nor will there be any discussion about how the waste might have been kept in sites from which it would be retrievable, on the assumption (or wish) that future technological developments may bring other possibilities.[71] Indeed the initial estimates of retrievability (that materials would be removable for twenty-five years) have proven to be gross overestimates. In many cases, the rate of deformation of the underground panels (three or four inches a year) exceeds technical estimates, effectively entombing the waste much sooner than anticipated. The material remains *retrievable* in a technical sense—it may be mined from the ground by whatever mining technology is available—but this is to stretch the meaning of retrievability beyond its intended sense.

As it now stands, three designs have been evaluated. One, Design A, was a variant of the trefoil design (Figure 10a), and another, Design C, was a variant of the menacing earthworks design. The rationale for not choosing either of the designs is given as follows:

> The quantity of material and general configuration of the berms give rise to a significant construction effort in their erection. For total quantity of material required, designs A and C each represent on the order of 1,400,000 cubic meters. Design B is approximately 750,000 cubic meters. In addition, the shape of the various berm sections for design C add an additional degree of construction complexity over that of designs A and B. Although design C is more "menacing," the actual warning of danger is conveyed effectively by the inscribed information on the monuments.[72]

Thus, there is a significant shift away from the conceptual problems of marking, toward a much simpler pragmatics of design and construction.

The semiotics of the site itself, however inadequate that discussion may have been, is passed over in favor of textual and pictorial inscription on the individual monuments.

> The primary purpose of the berm is to convey the Level I message that something manmade is here. All of the berm configurations will perform this function. Design A does not provide the degree of "protection" (i.e., enclose the repository footprint) that is conveyed by either design B or C. It is acknowledged that access to much of the footprint is inhibited by an additional 10 meters of material when design A is considered. However, other than causing some additional effort to set up a drilling platform on the design A berm, it adds little when considering that the repository is 655 meters below the surface. The volume of material required to construct the Trefoil shaped berm is considerably more than that required to construct design B. In addition, design A would not provide the same degree of protection from wind-driven erosion of the monuments as does design B. The proximity of the monuments to the berm in design B will provide more protection to at least one face of a monument than be available to the more exposed monuments in design A. Although barriers can be erected to improve protection

of the inscribed material there is no apparent advantage of the Trefoil over that of the perimeter berm. The elevated location of the Information Center at the center of the design A berm will also be subjected to greater wind-driven erosion effects than the more protected location provided by design B berm. The berm aspect of the three permanent marking concepts considered is the major design variable. The Monuments, the Information Center, the Storage Rooms, and the Subsurface Warning markers will not significantly vary in cost for any of the three configurations. When all the salient features including total materials required, ease of construction, meeting design requirements/criteria, and establishing permanence are compared, the conceptual configuration using a rectangular berm to enclose the entire repository footprint is the most practicable. For this reason, Concept B is the configuration of choice for the Permanent Marker System.[73]

The DOE contends that the assurance criteria can be met with the following design components:

> A controlled area of 41 square kilometers. About this perimeter of this controlled area will be 32 monuments, placed 805 m apart.[74] Each monument will be a two-piece, one meter square, granite monolith. And each will extend 5 meters below the surface, and 7 meters above ground. Each of the four sides of the monolith will be inscribed—both above and below ground—[with a star azimuth map similar to Figure 8], a pictograph sequence [Figure 16], and a warning message inscription [Figure 15].
>
> An Earthen Berm configuration will enclose the entire repository footprint (roughly 870m by 720m). The berm will be 30 meters at its base, tapering to a 4 meter flat surface, with an elevation of 10 meters. A number of materials will be used for construction. The surface soil will be excavated to a level of subsurface claiche soil (about 3 meters). A core will be built up using salt left over from the excavation of the repository. On top of this, a 2–3 meter compacted layer of claiche soil, then a meter of riprap, then a meter of a soil riprap mixture.

Large strontium ferrite permanent magnets buried within the berm at intervals of 75 to 100 meters will be used to give the berm a distinct magnetic signature.

> Trihedral shaped metallic objects will be emplaced at random to give the area an anomalous radar signature.
>
> Small buried warning markers—23 centimeter disks made of granite, aluminum oxide, and fired clay—will be placed at random depths and intervals throughout the berm area.

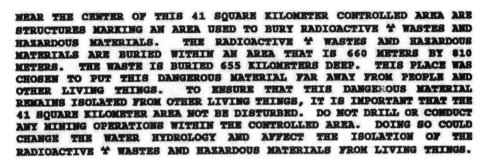

THESE MONUMENTS OUTLINE A
CONTROLLED AREA OF 41 SQUARE KILOMETERS NEAR
THE CENTER OF WHICH RADIOACTIVE ☩ WASTE IS BURIED

NEAR THE CENTER OF THIS 41 SQUARE KILOMETER CONTROLLED AREA ARE
STRUCTURES MARKING AN AREA USED TO BURY RADIOACTIVE ☩ WASTES AND
HAZARDOUS MATERIALS.   THE RADIOACTIVE ☩ WASTES AND HAZARDOUS
MATERIALS ARE BURIED WITHIN AN AREA THAT IS 660 METERS BY 810
METERS. THE WASTE IS BURIED 655 KILOMETERS DEEP. THIS PLACE WAS
CHOSEN TO PUT THIS DANGEROUS MATERIAL FAR AWAY FROM PEOPLE AND
OTHER LIVING THINGS.   TO ENSURE THAT THIS DANGEROUS MATERIAL
REMAINS ISOLATED FROM OTHER LIVING THINGS, IT IS IMPORTANT THAT THE
41 SQUARE KILOMETER AREA NOT BE DISTURBED. DO NOT DRILL OR CONDUCT
ANY MINING OPERATIONS WITHIN THE CONTROLLED AREA. DOING SO COULD
CHANGE THE WATER HYDROLOGY AND AFFECT THE ISOLATION OF THE
RADIOACTIVE ☩ WASTES AND HAZARDOUS MATERIALS FROM LIVING THINGS.

DO NOT DRILL HERE. DO NOT DIG OR CONDUCT MINING OPERATIONS WITHIN
THE 41 KILOMETER CONTROLLED AREA. DO NOT DO ANYTHING THAT MIGHT
DISTURB THE WATER HYDROLOGY WITHIN THE CONTROLLED AREA.

DO NOT DESTROY THIS OR ANY OTHER MARKER. THIS MARKING SYSTEM HAS
BEEN DESIGNED TO LAST 10,000 YEARS. IF THE MARKER IS DIFFICULT TO
READ, ADD NEW MARKERS COMPOSED OF LONGER-LASTING MATERIALS AND COPY
THIS MESSAGE IN YOUR LANGUAGE ONTO THEM.

FOR MORE INFORMATION, GO TO THE BUILDING NEAR THE CENTER OF THIS
MARKED AREA. THE SITE WAS KNOWN AS THE WIPP (WASTE ISOLATION PILOT
PLANT) SITE WHEN IT WAS CLOSED IN 2030 A.D.

Figure 15. Controlled area perimeter monument inscription message. Reproduced from Department of Energy, Compliance Certification Application, Appendix 3.

Within the repository footprint berm area, granite monuments—constructed identically to the controlled area monuments—will be spaced evenly (150 meters apart) within the perimeter.... Each meter-square granite monolith will stand approximately 7 meters above ground, and extend 5 meters below the surface....[75] As with the controlled area monuments, the four sides of the monuments will be inscribed—above and below ground—with two pictographs...—and each will be inscribed with the level II and III messages in seven languages, the six official United Nations languages (English, French, Spanish, Chinese, Russian, and Arabic) and Navajo.

A granite information center will be located at the precise center of the berm area. This will be a kind of kiosk containing all of the information presented elsewhere on the site.

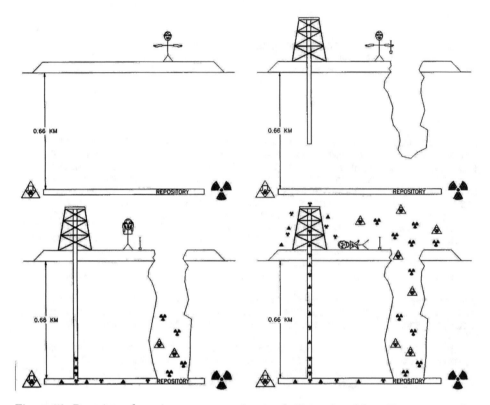

Figure 16. Repository footprint monument pictograph. Reproduced from Department of Energy, Compliance Certification Application, Figure IV-3.

A single building from the existing WIPP facility, the hot cell, located 320 meters north of the berm, will be left standing as an *archaeological artifact*.

A buried storage room will be located 6 meters below the surface, 160 meters north of the berm on a line passing through the information center, the center of the northern and southern sections of the berm, and the hot cell concrete artifact. An identical storage room will be constructed inside the berm itself. The location of both of these underground chambers will be documented off-site.

In the language of the Markers Panel, the site itself retains the quality of Level I message. That is, the berm and surface structures delineating the controlled area boundary and the repository footprint boundary adhere in presenting a sign that a man-made production is there. They write:

The monuments, information center, and buried Storage Rooms provide the surfaces upon which to engrave the Level II, III, and IV messages. The Level I message includes the earthen berm, the granite monuments, and the information center. The physical size of these structures should clearly convey the notion that the marker system is a manmade facility which required a significant amount of effort to construct. . . . This should provide the inspiration for any organization with sufficient resources to dismantle the surface structures to investigate and attempt to understand the purpose of the site prior to initiating activities which are counter to maintaining the site's integrity. Individuals intent on vandalism or artifact collection may cause some superficial damage. However, due to the size of the structures and the physical attributes of granite, it is very doubtful that they could significantly reduce the structures sufficiently to destroy the implication that something manmade occupies the site.[76]

The EPA rendered its final rule in May 1998 and indicated that it had been persuaded that the question of passive institutional controls, in particular the marking system outlined by the DOE, was adequate. The discussion of the question of meaning had been successfully shifted to a concern about the durability of materials.

For now the project is about transportation. Now that the WIPP is operating, all of the relevant sites and all of the transportation routes linking the country from four termini—Washington, California, Illinois, and Georgia—are linked into a disposal machine for the 38,000 shipments that will take place over the next decades.[77]

And here, the story of the marker is awaiting a form of resolution that can only be the passage of time. So far, the DOE has promised to undertake the following activities over the next century:

Design and test marker concepts and materials—1996–2083 (87 years).
Construct test berm—1998–2005 (7 years).
Monitor performance of test markers and berm—2005–2083 (78 years).
Test comprehension [by which is meant 'legibility'] of marker messages—2018–2023 (5 years).
Develop final design of markers—2083–2090 (7 years) [it is not clear why it takes as long to develop the final system as it does to construct the test berm].
Construct all markers (this phase presumably incorporates all quarrying, fabrication, and emplacement)—2090–2093 (3 years).[78]

Yet the DOE is now in an interesting position. The EPA's regulatory control of the WIPP extends only until the end of the facility's operation life—until the site is full, say 40 years from now. Consequently, any commitments made by the DOE for actions to take place *after* closure of the site are, in principle, unenforceable.

Perhaps none of this is all that surprising. A berm, some monuments, some signs. A generation down the road, most of the opposition to the project will be dead and also buried.

I have been trying to show how the conception of the problem as a technical and pragmatic question of security has led to an equally technical solution of burial and a monument. The question now becomes—in light of what has taken place—how do we read this response? How can we understand this response where signs of "enduring significance" will be enduringly significant without anyone there to point this out? Where there will be no grief, and no mourning; where meaning will transpire without us; and where there will simply be an installation that must be read correctly.

Threat and Trauma

Threat and Trauma

The Question of the Monument

MUCH OF the work of signing the waste has been directed toward the future, toward an event as though it is yet to take place. The desert monument as archive is a passive sender of information directed at future persons on a need-to-know basis. In this sense, the designs are not about the present, they are not about material that is dangerous now in the present and that has been dangerous since coming into being. And significantly, these designs disavow the assertion of their own presence as a massive production in the desert. All of this allows the monument to be placed, as it were, facing away from us, facing always toward the future as though it was not intended to recall any past in particular. The future witness to the monument is called upon to understand only the site itself—not the reason that the wastes are there, not the reason that the wastes came to be wastes, and certainly not any sense of how we as de facto authors and custodians of this waste might feel about having been responsible in various measure for producing it. Nothing, in other words, that might convey the basic truth that were we to reflect upon it, we could only feel a profound shame and sorrow with respect to a toxified present *and* future. What remains at a distance in all of this is that the disaster is not in the future; it has already happened (and, to paraphrase Caruth, we have survived without knowing it). The monument can only feign its prophylaxis, feign the impartiality of a description when perhaps something more in the shape of a confession is called for.[1]

So what is being asked here? What is this crazy plan in the desert? If we consider the WIPP problem as being really about making a monument that can endure time, in addition to having badly misconstrued the question, we have perhaps unwittingly (or otherwise) produced only a problem of a technical sort; a problem that is (at least in principle) solvable. That is, with the application of better design criteria, more enduring materials, and better models of future environmental conditions, a "better" (read: longer lived) monument may be produced. If we consider the problem to be one of making a monument not only endure but "mean" for the prescribed period, we have an utterly different sort of problem. In fact, we have not one but two problems side by side.

Let us say that if the first problem concerns meaning and its projection into the future, then the second must concern the vehicle of this transmission; that is, the second problem is the monument itself. And once it is put this way one might begin to ask certain questions concerning meaning and monuments. For instance, is the monument being demanded for the nuclear waste burial even the sort of task that monuments are called on to perform? Perhaps. This monument, such as it is, is being called upon to stand for a sort of remembrance. Not unlike other monuments, war memorials, and the like, this monument would be responsible to history as a reminder for us. And even if this makes sense, what monuments are conventionally charged to call into remembrance is something for which we wish to be remembered. "A work intended to celebrate and preserve the memory of a person, an event, or an idea."[2] The great battles, the great figures in history, moments in time, points in space. Borglum's Rushmore.

One could say that a monument *to* something is an anchor of presence dropped into time by a people unsure that they will be remembered or perhaps *how* they will be remembered if they are at all. Monuments are left to posterity to things worth remembering, to things of value, that we value, things for which we wish to be remembered. In this sense, such undertakings are not exactly about the future. Rather, they are about the anxiety of the present—an ontological anxiety— precisely with respect to the very uncertainty of the future *(le dur désire de durer)*. The desire is to make permanent that which threatens to disappear irretrievably. The very idea of a monument to something that we wish would never have come to presence to begin with—and something that persists (literally) in the present, and actively performs its danger on the safety of the future, even as it impinges on our own—is a very unsettling thing.

I do not mean to say that monuments are always affirmative, that the function of a monument is always to commend upon positivity. There is also a deep affinity between monument and atrocity, between monument and disaster suffered upon community and memory. And indeed the late-twentieth-century countermonument may be read as having as its principal concern the very anxiety of its own assertion as a monument.[3] No longer is the public space of the monument an unproblematic site of memory, a prophylactic against forgetting, and compensation for the symbolic debt of the past. The commemorative responsibility of the modern monument (regardless of what it is or how well this responsibility may be met) is to *stand for* events. It is the monument that is the bearer of the responsibility—not a community and not memory—and this is precisely what countermonumental strategies seek to challenge. The countermonument seeks only the afterimage of its effect, the memory, the remainder of the monument's assertion. As a political praxis, and an ethical injunction, the contemporary ethico-political shift in the practice of the monument is a troubling development for the (strictly speaking) modern monument in the desert. Monuments that disappear, that inhabit the negative space of their shadows, that are assertions of an absence (present and otherwise)—these countermonumental features simply add to the ambiguity of the desert project.

It is, I think, important to remain aware that the commemorative work of such monuments—even if the injunction "Do not forget" remains the same—is always a function of the particular needs (political or otherwise) of a given community. One can see this in the way that Holocaust memory is commemorated in remarkably different forms according to the location and time in which it is carried out, the political climate in which it is constructed, and the specific character of the rupture with the past it attempts to bridge (e.g., the victims, the survivors, their children, the perpetrators, the liberators).[4] The way in which monuments mean, or refer, is obviously never straightforward. The didactic, symbolic, and functional dimensions of the monument must always come into conflict with other countervailing (or potentiating) forces: denial, forgetting, disavowal, and resistance. This is not to say that some monuments do not favor certain forms of historical memory over others (official vs. community sites, community vs. individual mourning). But there remains a necessarily participatorial dimension of the monument as site. The place, that is, where the intersection of the monument and the witness constitutes an event, the place where one may locate the production of a meaning.

It is in this sense, precisely, that a monument to waste is an inversion of the work of a monument. To the extent that it draws our attention it does so not exactly to the past, nor exactly to the future. Its task is not to perpetuate memory, nor is it a device for recollection. It exalts nothing. It must make an assertion—that of its danger—but it must do so in an idiom, if not entirely foreign to such messages, then at least one to which it is very ambiguously related. Its concern is the perpetual present, a *now* that is indifferent to history, yet one in which the witness must not fail to understand.

The official position on the desert monument acknowledged the difficulty of the projection of meaning, the fragility of memory, and the unprecedented temporal duration of concern. Yet, the recourse to a system of monuments to deal with these problems was deemed more or less obvious. Some thought was given to the persistence and intelligibility of various human-made monuments (e.g., the pyramids of Giza, 2600–2500 BC; Stonehenge, 2700–2500 BC; Nazca Lines of Peru, 200 BC–AD 600; Serpent Mound of Ohio, 1000 BC–AD 700; the Acropolis, 447–424 BC).[5] And as discussed above, some thought was given to the temporal dimensions of communicative acts, the structure of future societies, and the shifting landscape of language. The problem, however, is that the monument, wedded to its site (as though it were a ruins) is not envisioned as equally wedded to (any) memory. In this sense the monument is seen as archival. The site will disgorge its meaning to its witness merely in the form of a warning with respect to the site itself. In this way the monument's concern in its gesture toward itself is that it convey a mimetic performance of the danger of the waste, the length of time it has been interred, and some manner of conveying a "do not open before" date. That is, a message, not a memory. A message that would alert us to the presence of threat.

So to all of the other things of which we prefer not to speak (and therefore speak about ceaselessly)—death, for example, or madness—we must add another category—ecological or nuclear threat. This kind of threat is threat of an ontological character. But this doesn't mean that it is really big, that because it is ecological it is about its big scale. It is not just about size; it may well be vast, but this is not the important feature. It is also that it operates both above and below a threshold. Ecological or nuclear threat cannot be understood as a simple threat, understandable as a promise or debt, and accountable in terms of risk and reparation. It is more complex, diffuse than this, and as such presents a problematic that exceeds traditional (or at least conventional) modes of conceptualization. This sounds more obscure than it is.

Industrial threats (as opposed to ecological threats as we mean them here) could be measured, quantified, or at least were in principle amenable and adequately expressed by concepts of scale. A collapse of a mine shaft—horrendous as it could be—can produce only so much effect in a community, on a people. It would make no sense for a survivor to wonder if they are actually a casualty (in a significant sense—I don't mean whether one's life was touched by a death). With ecological threats, the question, Am I already a casualty? is exactly the question provoked. But even the concept of the casualty undergoes a transformation. Casualty regains some of its archaic sense of "chance" and "fall" (in the sense of "you just don't know"), but also takes on a quality of undecidability and indifference. Am I already a casualty? is a senseless question in the context of a train wreck that today makes perfect sense. An ecological casualty can be an organism that is only a vector for damage directed at a subsequent generation. And in any case, whether one is or is not a casualty is not the point. The point is whether one suspects one is, since it is very possible for there to be no perceptible difference.

It is as though our senses, our very perception, had been expropriated, rendered useless and vestigial in the face of threats that cannot be seen, heard, smelled, tasted, or touched.[6] The appeal to the eyewitness (even one's own eyes) comes to have little value here. There is nothing there, nothing to be seen, leaving us dependent on others (often the same others, that is, the same institutions that produced the threats) to determine the appropriate means (instrumentation) with which to represent it back to us and for us. Much the same can be said for concepts such as safety and hazard. What is dangerous and what is safe, what dosage is hazardous and what is not, such thresholds and limits obscure the fact that they are foremost creatures of politics and not the test tube, objects of persuasion, not measurement. And this, I would add, is exactly what people like Kai Erikson and Robert J. Lifton will say is the profoundly damaging—that is, psychically damaging—dimension of nuclear and toxic threats. We just don't expect to be injured that way. If we're hurt, we expect to know it, and to know why; the right to own one's pain. And if something happens, we expect to know when it began, and when it ended.

So then perhaps the question becomes, How do we conceptualize such threats as are presented by nuclear and toxic and other ecological events? The project in the desert demonstrates one set of contortions that a government has gone through to respond to such threats. For our purposes here let us say that we have several things that need to be explored. The first is how to think about the social response to this kind of threat. After all, in a way that's all there is to see. That is, a bunch of responses—some very calculated, some not at all—in relation to something that may not even be visible or tangibly present. To put this differently: How might we come to think about the psychology of threat?

The second question then falls on the other side, so to speak. It has to do with thinking about the threat itself. How do we represent (talk about, understand) ecological threats of the sort we see being interred in the New Mexican desert without reducing them to the phantasm of a psychological event? Obviously both of these questions are very difficult, are mutually implicated, and indeed are hard to maintain as distinct questions at all.

Trouble with Risk

The modern idea of risk, says François Ewald, is but a neologism of the insurance industry. And insurance itself can only be understood as a technology of risk, designating not so much an event as a manner of thinking about an event's possibility in terms of a probability. And in doing so, in its determination of a manner of fixing a probability, insurance can then maintain itself as a profitable way of distributing loss, such that the loss is felt less by those who are injured.

> Rather than with the notions of danger and peril, the notion of risk goes together with those of chance, hazard, probability, eventuality or randomness on the one hand, and those of loss or damage on the other—*the two series coming together in the notion of accident.* One insures against accident, against the probability of loss of some good. Insurance, through the category of risk, objectifies every event as an accident. Insurance's general model is the game of chance: a risk, an accident comes up like a roulette number, a card pulled out of the pack. With insurance gaming becomes a symbol of the world.[7]

And it is a large and unwieldy game indeed. A decade ago the insurance industry spoke of potential losses from a single massive catastrophic event ("supercat events," as they call them) on the order of $100 billion—a single event that would be four times the total claim amount for the twelve years between 1980 and 1992. The magnitude of losses progresses arithmetically (if not geometrically) with each major event. The year 1994 brought the Northridge earthquake in Southern California, the most recent costliest disaster in American history (estimates of $44 billion), surpassing the previous most costly event, the 1992 Hurricane Andrew ($26.5 billion). In 1995 an earthquake in Kobe, Japan, caused more than 6000 fatalities and over $100 billion in economic losses. The earthquake in Izmit, Turkey, in August

1999 caused 20,000 fatalities and an estimated $20 billion in economic loss—about 10 percent of that country's gross domestic product.[8]

Recent trends in seismic science figure that the tectonic deficit being accrued in Southern California could well result in a cascade of earthquakes where dollar loses would be in the trillions.[9] The only way for the insurance industry to see potential amid the prospect of actuarial apocalypse is by insuring itself, by becoming part of the game of chance. Thus, "reinsurance" becomes part of the industry. At this level reinsurance amounts to speculating on disaster futures, and in the absence of (or in advance of) the "big one" there is a lot of money to be made. However, reinsurance can only partially cover this increased exposure for the insurance industry, in part because the reinsurance industry is not capitalized enough to handle a single massive supercat event and in part because insurers tend to reinsure for lower levels of loss than would occur with a supercat incident. (Reinsurance is sold in layers of exposure above a certain deductible or retention level, e.g., between 100 and 200 million.) But the cost to insure for higher levels of exposure (in effect paying very high premiums for low probability supercat risks), even if underwriters are prepared to offer the coverage, is prohibitive. There is a suspicion in the industry that if insurers can pass on risk to reinsurers, they will do so in a way that passes on the least attractive areas of risk; the industry calls this a "moral hazard." Others would call it capitalism. And if you live in New Mexico you might say that the concept of moral hazard explains why the WIPP is there to begin with. The result in any case is that the risk of catastrophic events is not efficiently redistributed, leaving insurers, the insured, and ultimately those not insured, in a perilous situation.[10] In the face of events of such magnitude—massive quakes in high population areas, tropical storms, toxic events, nuclear events—the concept of the insurable breaks down. Or, more to the point, it already has and the industry has known this for quite some time. Indeed, with specific reference to the nuclear catastrophe (in the United States), the question of the insurable was largely dismissed with the 1957 Price-Anderson Act that capped the liability limits for accidents occurring at nuclear electric utilities. This creates a strange loop in the nuclear industry's position that a large-scale accident is only a minuscule probability—simultaneously diminishing industry credibility and legitimizing the "environmental" position that industry characterizes as irrational.

Modern threat—or modern "ecological risks," as Ewald calls them—is unique. (I make this claim repeatedly because the contrary belief is so profoundly knitted into contemporary thought.) To clearly grasp this as more than intuitively true, it is necessary to consider both the transformations wrought by threat or ecological risks and the lack of *fit* between these transformations and the traditional means for thinking about such things. The "new generation" of risks exist in an uneasy relation with the traditional conception of responsibility. Such risks share the following characteristics:

> In terms of potential damage that has to be covered by insurance, they are on the level of natural catastrophes. They concern entire populations, whose withdrawal, removal, or exodus must be planned for (Seveso, Three Mile Island). They are on the order of a disaster. Unlike an earthquake, however, they derive from human activity, from technological progress, and as such are if not known then at least foreseeable, extrapolatable, and accepted: they are artificial catastrophes... they do not concern individuals taken separately... so much as the biological balances between a population and its environment.[11]

The uniqueness of contemporary ecological threat is also seen in the manner in which it is distributed both above and below thresholds. Above, in the sense of the transnational (and transgenerational) character of a Chernobyl. And below, in the sense of risks that operate at and below the level of biology. In the former sense, the effect is direct, but simply too large and too complex to be dealt with in terms of a model of responsibility. For global warming, or ozone holes, or aquifer depletion, there is no responsible party, no place where in the final instance the buck would stop, for two reasons. First, it is not simply because of complexity—as though it were a problem of sorting out degrees of responsibility, of culpability, of foresight, and so on. It comes down to the very well-worn concepts of cause and (reasonable) doubt. Second, from the point of view of the insurable, responsibility belongs to a juridical logic (reliability, judgment) and not to risk logic. Risk logic calculates only the probability of an event taking place, the probability of the accident. This is what gives it its appearance of objectivity.[12]

The calculability of risk presumes that whatever event it is that we wish to speak of can be made part of an actuarial calculation. With ecological risks, risk—because nonlocalizable and noncalculable—becomes generalized and no longer a matter for insurance.

Within the second aspect of threat—the sense in which such threats operate below a threshold—the risks become insidious. Virtual objects. And even though it may represent a threat to continued life, it is again significantly nonlocalizable and, as such, is displaced in relation to a victim/perpetrator model. In both movements of ecological threat, there is a convergence on a point that is purely excessive. Not the body-as-capital, for which compensation is possible, because an equivalent is (by definition) calculable—as with traditional insurance—but the very ontological status of the body-in-environment.

Given that ecological threats result from human (technological) activity and that they are manifest on the order of what is traditionally thought of as a natural disaster (earthquake, volcano, mud slide, typhoon)—artificial catastrophes, as Ewald puts it[13]—it is not surprising that they are thought of as only quantitatively different than traditional threats. Such events can be only partially understood in terms of cost/benefit analysis. Traditionally we would say that risk is the term that mediates between cost and benefit. Risk, above all, is calculable (and it is tautology to say so). Thus, to speak of ecological risks we must also allow that there exists some degree of objective ground upon which to make a comparison between costs and benefits, some way to provide a fulcrum between the two terms. But to evaluate a cost, one must have a conceptual grasp of its lineaments; one must be able to point to an activity and say *this exceeds*, in its distributive character, *any possible benefit*. But no such judgment can be made on the order of costs and benefits. Such judgments operate on incommensurables. Effects can exceed causes temporally, spatially, and proportionally. Unmoored from these coordinates, effects may and often do run counter to the initial, stated, or implicit intention—the Green Revolution, thalidomide, tainted transfusion blood (e.g., hepatitis C). Causally, the picture can exceed any standard account of a forensic accounting, not simply a complex and cascading causality in which causes cause causes to cause causes (Wilden)—which ultimately remains a (big) causality in linear mode—but a situation in which causality as such becomes less important than the ability to creatively intervene within and among effects. Far from the Darwinian fatalism of Hardin's *nature will commensurate the incommensurables* (which tacitly gives way to the instrumentality of technological demands[14]), *ecological threats can find a meaning only within the social.*

Ewald sees in ecological threats first and foremost a threat to democracy and any idea of a fundamental social contract. No longer do threats turn on distinctions of public versus private interest. "Ecological risk divides society against itself at it most intangible, least measurable, and perhaps most essential point: it divides society on what is supposed to unite it, on its values, on the definition of its collective interest."[15]

We cannot, it seems, decide among ourselves what sorts of risks are worth taking nor how we might go about making such decisions. Which is not to say that ecological risk, and the social rift that it opens, does not in fact speak to us. It does, because in it we speak to ourselves. What it does not disclose though, even under the most "objective" of conditions, is anything at all to do with limits and propriety.

There is nothing objective whatsoever about the technologies of risk. It *tells* us noth-ing. It is a "category of understanding; it cannot be given in sensibility or intu-ition."[16] It is manifest only insofar as a group elects to allow its existence. Risk be-comes acceptable simply through the paradox (tautology) of ecological threat; that is, the

> bigger the objective risk (for example, one on the scale of a catastrophe), the more dependent its reality is on a system of values.[17]

Think about this. It is not difficult to see a relationship between risk and a social perception of value. The magnitude of the risk is in a sense acknowl-edged by the magnitude of the response. To return to the WIPP, the fact that we wish to dispose of the problem through a burial without mourning suggests that the granted "reality" is only sufficient for it to be laid to rest. The dose and the thresh-old—and the wish—here are implicitly reflected in the response, in effigy. Which is to say two things. First, there is a quantitative and qualitative arbitrariness to the understanding or perception of large-scale risk. It draws its motivation not from an objective well of nature, but from *our* labors of understanding, our labors of making sense of the world. And second, it is nonetheless perfectly real (in a completely con-ventional sense).

Risks become good or bad, acceptable or unacceptable, for rea-sons that have to do with politics, not nature. Which is not to say that there are no risks—risks are everywhere but (and this is what Ewald has said) once confronted by threats of sufficient magnitude, there is no outside "reality" to which one may ap-peal. Any appeal to nature as an arbiter in disputes about risk is operationally point-less and politically foolish. The implication is that there is no justification to assume that in ecological risk one finds a firm objective grounding upon which to contest political or economic practices.

> With ecological risk, nature becomes social through and through; the prob-lematic of nature is overtaken by radical artifice. The ineluctable conclusion of the logic of balance: everything becomes political, down to what seems the most natural in nature.[18]

Nature is thus not a well of true speech, uttering a language for-eign and forgotten. It is simply a manner of concealing the fact that we are con-demned to live in an order of pure politics, and pure decision. A transformation takes place in the becoming-political of nature (or the becoming-natural of politics) whereby death is no longer situated beyond the edge of life. If anything, such risks as ecological threats resituate death into life in the form of an unknowable risk. And for Ewald, the realization of this new relation between life and death can either give way to an anxiety, to a kind of "collective and individual frenzy of self-protection" and "denial behavior," as he put it, or—and this is an important utopic moment—it can also give rise to new forms of life, to new intensities, to transformations: perhaps a manner of life *with death* that would radically and revolutionarily transform new kinds of subjects. This is a provocative notion. Ewald's example: the "American survivalists." He sees them as constituting a manner of being stronger than death, as an attempt at a conquest of a new identity. Survivalists as the proletariat of threat? Not likely. After all, the real refuge of survivalists is founded by and accommodated within an idea of the "natural." Yet in a certain sense survivalists have undertaken a radical decoding and movement on the level of particular aspects of the social. But it is also certain that they retain and deepen territorial linkages with some decidedly historical principles of righteousness, autonomy, and freedom, etc.[19]

What is missed here is the nuance of ecological threat as a demand for a response. "Against the ruin of the world," as Rexroth put it, "there is only one defense—the creative act."[20] It is important to resist drawing the conclusion that since any response can only be political (read, subjective) and that under such conditions life is radically altered, therefore the only response operates in the service of survival.

Society of Risk

A head of state representing the forty-two-nation Alliance of Small Island States (AOSIS) stands before Western and Eastern world leaders at the 1997 Kyoto Climate Treaty Conference. He declares that unless global warming is halted, the homeland of his people will disappear under the rising Pacific Ocean. "Ignoring our pleas will amount to nothing less than denial of our rights to exist as part of the global society and of the human race," he said.[21] A staggering claim and one that in a significant sense cannot be challenged—the water is rising. It matters little that we might agree upon a causal picture. The conditions within which the future can be thought for these small island states, perched as they are a couple of meters above sea level, has been radically altered. The future is now organized around the discourse of threat and disaster. Not simply game over, but the game's rules, its history, the memory of it having been played, and its culture.[22]

In October 1999 the "super cyclone" 05B, as it was called—the most intense cyclone to hit India in that century—having gathered intensity in the Bay of Bengal for several days, finally slammed ashore at Paradip in the state of

Orissa. Although the storm destroyed all the monitoring equipment in its path, wind speeds were likely on the order of 300 km/h. Some 500 km of coastline was flooded (up to 50 km inland), 15,000 people drowned, and more than two million people were left homeless.

However, unlike the previous five cyclones that hit the region in the six months prior, this storm was exceedingly slow, around 15 km/h—not much faster than a bicycle might travel.[23] This ironically slow approach should have provided an opportunity for evacuation. The meteorological office in Calcutta had been advised of the cyclone and its increasing magnitude two full days prior to the coastal regions being notified, yet the coastal regions were given only one day's notice. Reports conflict, but it seems that the satellite-based cyclone warning system (a network of sirens) did not function. Instead the region was submitted to the full magnitude of the storm for nearly 24 hours. The tragedy here cannot be entirely understood as issuing from a hostile nature, nor merely as the vulnerability of humans, nor indeed as the expression of complexity (the traditional trinity of sociological disaster models). The disaster is an expression of a host of factors, many of which have not the slightest to do with wind or waves. And in its wake, it is precisely the failure of warning and response mechanisms, not the storm and most certainly not nature, that are considered to be answerable for the disaster.

Where might we locate nature in this? What is referred to by something on the order of global warming is if nothing else the contested product of human agency, more specifically of twentieth-century technoscientific and industrial practice. Through large-scale hazards on the order of global warming or nuclear threats, nature (spun increasingly through technoscientific debates) is revealed as a field that is irreducibly political. These hazards, these technoindustrial productions, become objects of politics not nature.

In other words, to say that threats too are politically constituted should not be controversial. Take, for example, beautiful Southern California. Until quite recently, there were officially no tornadoes. It turns out, however, that there are nearly twice as many annual tornadoes as there are in Oklahoma City. Tornadoes are bad for business, so the "secret Kansas," as Mike Davis calls it, decided not to have any.[24] And the only way to do this (in light of the periodic funnel clouds and high winds and other tornado-ish features) was to not talk about it.

Such is the society of risk, as sociologist Ulrich Beck calls it. The example of contemporary California differs from examples of early industrial configurations in that there is no such thing as either the promise or the fact of an ecological proletariat (survivalist or otherwise). This is not to deny the existence of

South Central LA or the distinctions between it and the pool-owning, gate-building residents of the Malibu hills. The point is that ecological threats do not divide neatly within these boundaries, nor do the boundaries (such as they are) prevent unfore-seeable transactions and surprising exchanges from taking place. That is, ecological threats construct a social cartography that is often, and largely, foreign to divisions such as class, property, and distribution and have a propensity to cut through social divisions, assembling new lines of affinity, new constituencies of those at risk.

Quips Ewald, threat is "paradoxically, creative!" Take Tuvalu, a nation of nine very small islands in the South Pacific (including the former British colony of the Ellice Islands). Slowly disappearing under the rising ocean, Tuvalu made a remarkable deal that would allow them to claim influence on the global political stage, while improving—if temporarily—their internal domestic situation in a stroke. They sold their World Wide Web domain name: .tv (dot tv) to a Canadian firm.[25] With the first installment of the sale (which is said to total $50 million over 10 years) they were able to pay membership fees to the United Nations as the 189th member State and have enough left over to put a large infusion of cash into health care and infrastructure.[26]

In any case, we could say, following Beck, that the distinction between risk and threat is the expression of a fault line between industrial and mod-ern societies. In the former, conflicts turned on the distribution of "positive" value: profits, prosperity, progress and its promise. There was something at stake for which a decision, a trade-off, could be made. On one hand, and on the other. Thus the negative side of the equation could be conceptualized as risk, precisely because the positive side could be readily identified and evaluated. In modern societies, however, risk is subsumed and transformed by threat because contemporary technological practices embody externalities that exceed both social and temporal limits (both above and below thresholds). They exceed limits of accountability as well, and in light of this, fail to fit any standard notion of compensation. It is a game between "losers, who refuse to admit the damage, who shrug it off, and repress it."[27] And fur-ther, this is the reason why threat and knowledge of it are so difficult to disentangle. Again, risk is calculable (by definition, and therefore arguably), while threat, on the other hand, is not.

From this point of view, the presence of modern threat is in no meaningful way an environmental problem. It is, rather, an institutional crisis. Threats are slippery things. They are, observes Beck with concision, "produced industrially, externalized economically, individualized juridically, legitimized scientifically, and minimized politically."[28] And in the public consciousness, the surplus of possible threats allows for easy substitution, modification, and transposition. "Just in time" threats. If air pollution from coal-fired energy production is the threat du jour, or if it is simply the volatility of Middle Eastern oil prices, nuclear power generation may reenter the market "defensively," through the back door of current anxiety and collective forgetting.[29] In the language of game theory, ecological threat is a negative-sum game of collective self-damage.[30] A global strategy asserts itself in the form of determining an equitable manner of distributing loss (i.e., negative conflict)—traditionally the role of various forms of social insurance.

So, to the claim that the bigger the objective risk the more dependent its reality is on a system of values, we now add Beck's corollary:

> Resistance to acknowledgment of threat grows in direct proportion to the threat's size and proximity.[31]

Disavowal and denial behavior vary in direct proportion to threat. It is not simply that threats of magnitude are dependent on "values," it's also that the greater the magnitude, the greater the resistance there is to even constituting them symbolically in any coherent fashion.[32] Threat is thus closed off from two directions, at least. The first is that the nonobjective status of threat requires acknowledgment—if not a consensus, then at least some form of social agreement that such and such impinges on someone's or something's welfare. Without at least this, there really is nothing to talk about. The second is a parallel movement that is itself a resistance to the very taking-place of social practices of acknowledgment, contestation, and agreement.

Whether we speak of disaster theory, or risk and insurance, threat can be understood—if at all—as a function of a probability, which, although common to the other conceptualizations, stands as a habit of thought in its own right. The question that probability poses—a probability invented by the same rule-governed, technoscientific endeavors by which the threats have been produced—is always the same, just its constituency changes.

Probability conceals (by rendering abstract) the fact that technological disaster is produced not only by technological design, but by (nonconsensual) social decision making. Thus, threat itself either is of no account or is of interest only insofar as it is a measure of something else (e.g., risk perception, social "irrationality," risk communication).

The WIPP project is replete with acknowledgment that it is not possible for it to do what it is supposed to do. That is, it is not possible to keep the wastes secure for even the legislated period of time. It's just too long. But what we have is a plan that looks like a solution even as it admits no solution is possible. As we have seen—whether we are speaking of active institutional controls, the security of salt formations, the concern for the transmission of "information," or the constitution of future societies—we keep running up against paradox and the very limits of "useful" speculation.

The Pathologies of Threat

When speaking about disasters, death and suffering, acts of terror, or any kind of limit experience, one is always at a loss for concepts, always struggling for words. *It was unspeakable, it was beyond comprehension.* In the face of the extreme, we as witnesses or survivors, or scholars or neighbors, become acutely aware of the limits of the expressible. One's capacity to articulate is revealed in its utter poverty in relation to an event. Accordingly, one may seek specialized discourses with which to speak about such things, to render them expressible, and to "communicate" them. Historically these have been such things as science and religion, poetry and painting, insurance and statistics. But psychoanalysis, which perhaps of all critical discourses— and in spite of anything else we may have to say about it—is optimized precisely for thinking about limit regions and crises in expression. It is also one of the most influential and tirelessly explored critical discourses of the late twentieth century. (Odd then, that to date, at least as far as I can tell, there have been no psychoanalysts or psychoanalytically informed theorists consulted by the DOE.) It is also interesting from the point of view that both psychoanalysis (Lacanian psychoanalysis) and the WIPP project share a similar obsession: signification. Mighty is the signifier.

The difficulty of threat is that the categories with which we attempt to understand it seem to operate either prior to or after the advent of threat itself—either symptom or pathology. But we have as yet no way to conceptualize the manner in which threat goes about its threatening. If not risk, that is, if not the possibility of something taking place or not, then what is threat?

We can say that threat performs itself, that is, it threatens. As threat, it is something that threatens to take place. The condition of threat is that it is always displaced in relation to itself; it is never fully contained as an abstraction— as a risk, or as a possibility. We can say that it is always partly in advance of itself. Partly, because it is not just that threat is notice of something that might take place; threat *is* something taking place. But threat's promise is not prognostic; it is not a sign or portent. What threat threatens does not resemble the threat. Nor is it stochastic, deriving a randomness solely as an expression (or inverse function) of a probability. (Although once it is disentangled from these confusions, it may be seen as socially heuristic.) It gathers its force by threatening to take place. Once realized, once threat makes good on its promise and takes place, is no longer threat, it has become something completely different. It becomes an event of some kind. In ceasing to be threat does it become something equally abstract? an accident, a disaster, a catastrophe? Or is the event anything but abstract? This depends on what we choose to focus on. So is there is a basic asymmetry between threat on one hand and the event on the other? Yes and no. Yes, in the sense just mentioned, that the event does not resemble the threat. And no, in the sense that in both cases we are left fumbling for concepts. So how to talk about threat?

  The psychoanalytic theory developed by Lacan is a very suggestive way to talk about all of this. The concepts of the imaginary, symbolic, and real—elaborated in considerable complexity over the course of his career—offer one way of positioning some of the difficulty of thinking about ecological threat and the events connected with it. While it is true that this trinity of concepts has had a lot of airtime in the past twenty years, I think there is a diagnostic value that will be worth our while to explore. First of all, they offer a potent language with which to speak about responses to the extreme. This is the value of psychoanalysis to begin with—it gives voice to the unspeakable: the real.

  The real's relationship to ecological threat is not fully explored until Žižek appears. In an otherwise playful exploration of Lacanian concepts set adrift in the sparkling filmic traditions of Hollywood, he chooses the "ecological crisis" to illustrate the workings of the real. For Žižek, the Lacanian real is a caution (or properly understood, injunction) against finding in ecological crises or threats a message or meaning. Foremost, the Lacanian real is senseless, meaningless. What Lacanian concepts offer, he says, toward an understanding of ecological crisis is "simply that we must learn to accept the real of the ecological crisis in its senseless actuality without charging it with some message or meaning."[33] It is here that Žižek makes an insight that puts Lacan firmly into debates about the cultural pathologies of ecological crisis and disaster in general. There are, he tells us, three typical responses to the threat of ecological crisis. All such responses are directed at doing one thing—they blind us to understanding the nonrepresentability of ecological threats; which is to say, they blind us to the real.

  By far the predominant reaction to ecological threats belongs to those who resist the very idea of a crisis. This operates in the register of disavowal (*Verleugnung*): *I know it's true* (whatever it is) *but all the same...* (I will behave as though it were not the case). Clearly, there are large doses of this at play with the WIPP project. For example, the project proponents acknowledge that the site cannot be made (and guaranteed) secure for the required 10,000 years. Nonetheless, they proceed as though it were possible. Or again, it is acknowledged that the science of radioactivity is beyond the comprehension of the average person, but nonetheless, the monument design details all presuppose the persistence of this very comprehension over thousands of years. Or perhaps more sweepingly, everyone involved seems to know that the WIPP will do little in the way of reducing the accumulations of transuranic nuclear wastes (and nothing at all about high-level waste), but nonetheless its accomplishment is spun as though it will.[34]

For those who respond to the threat of ecological crisis in a register other than disavowal, there are two modes. First of all, there are those who respond with obsessive (that is, neurotic) activity. Second, there are those who read into the crisis (which is to say, they project into it) a message or meaning.[35] Žižek takes these three responses—"a fetishistic split, and acknowledgment of the fact of the crisis that neutralizes its symbolic efficacy; the neurotic transformation of the crisis into a traumatic kernel; a psychotic projection of meaning into the real itself"—and organizes them as essentially means of—or strategies for—avoiding an encounter with the real of ecological threats.

In the first instance, the threat provokes an obsessional economy and frenzied activity aimed at preventing the calamitous X from taking place. It doesn't matter what X is; the point is that it must be prevented and to do so requires constant vigilance and activity. Julia Kristeva describes the activities of the obsessive as valuing the procedural over the declarative. The obsessional associates each situation with a requirement to *do* something—if *a*, then do *b* (not *a* means or explains *b*).[36] Kristeva calls this as a "paradoxical doing," an act minus one, because it is a kind of activity (doing) that is deprived of its affect. For the obsessive, the result is a compulsion to locate semiotic means of (displaced) expression—gestural, visual, or mobile.[37] This may have something very important to say around understanding certain responses to threat. Think of certain forms of obsessive activism (or kinds of activists), certain health practices, and so on. Indeed, the near magical construction of an organic body of the earth—for example, Gaia—is a particularly massive instance of such a displacement and procedural obsession. We could wonder whether a disruption and displacement of an affect at one level (for example, the social or the family) finds its way into an emergent caring and nurturing position with respect to the biotic, "Mother" earth.

The third response to ecological threats—or the second nondisavowal pathology (which resonates somewhat with the "nature bats last" school of environmentalism)[38]—is semiotic. Threat and crises are taken to be very specific kinds of "signs." Threat is not a command to *do* something, but rather a message to be understood and heeded. As a sign or set of signs, the ecological crisis is generally presumed to be indexically related to a normatively charged (and generally pissed-off) nature. The crises—global warming, ozone depletion, population, nuclear weapons, water, the Soviet nuclear industry, post-industrial Eastern Europe, smog, and AIDS, whatever—are read symptomatically as providing a link between a manifest crisis and a disrupted or transgressed nature. But foremost, these signs tend to tell a story— or at least are the implications of a larger narrative—concerning the ecological, and therefore moral, improprieties of "Man."

Problems with the Real

The signifier is stupid.

Lacan, *Encore*

These pathologies of response are united on the level that they are directed at blinding one to the fact of the "irreducible gap separating the real from the modes of its symbolization."[39]

But formally at least, the pathologies and the responses to ecological threat that constitute them speak only of reality, of responses in a lived world. From a psychoanalytic point of view, what about the ecological crisis that connects it with the capital R real? To begin with, one must not conflate the real with "reality." Reality is the job that you do, and the temperature outside, and the noisy neighbors. But this is not the real. Reality is everything that has already passed into a symbolic and imaginary matrix. Which is to say, everything that is constituted by our relations to language and to ourselves. The real is that which has not been symbolized. Moreover, it is the null point at which symbolization fails; it is independent of and indifferent to attempts to symbolize it. All the comfortable (or uncomfortable) divisions of reality have no place in the real. Up, down, here, there, inside and outside, these things make no sense in the real: "The real is without fissure."[40] Nor does the real know anything of lack. It is the beyond of the symbolic; we can have no knowledge of it. It is only discovered by the distortions it produces in the symbolic world, but in turn, the symbolic can only function by circulating about these zones of distortion, these hard places where symbolization falters. The real, that is, that "resists symbolization absolutely."[41] Even the expression of the events of chance—the throw of the dice, the random event—are prestructured and made sensible only within a symbolic framework.[42] "The very notion of probability and chance presupposed the introduction of a symbol into the real. In the real, at each go [throw of the dice], you have as many chances of winning or of losing as on the preceding go. This only begins to have meaning when you write a sign, as long as you are not there to write it, there is nothing that can be called a win. The pact of the game is essential to the reality of the experience sought after."[43] This is a problem. If there can be no knowledge of the real there ought to be nothing to say about it. We can say nothing at all about the real apart from the disturbances it happens to create in the symbolic fields which we inhabit. (The real does not help those wishing to be more proactive.) This is not at all hyperbole, though one must exercise due care. Identifying the real with ecological threat can—without care—divert attention from important projects of saying and knowing and in doing so can provoke us to increased levels of silence. In aligning threat with the real, even in a conjectural manner, we want to enable insight into threat, not to shut it down.

Chernobyl illustrates well the liminal characteristics of the real's irruption into the symbolic and into reality.[44] It is here that we see how the real shows up when something breaks down, in the very moment when something fails, the moment where there is a sudden and overwhelming evacuation of meaning. The Russian filmmaker Vladimir Shevchenko headed the first film crew that was permitted into the "red zone" (a 30 km^2 area that was emptied of 100,000 residents in the days and weeks following the accident). The short, part black and white, part color documentary that was produced, *Chernobyl: Chronicle of Difficult Weeks*, is in one sense simply a clumsy piece of back-slapping propaganda showing how well the Soviet scientific, technical, military, and Party authorities came together in the face of great adversity to overcome the severity of the accident. We see footage of many meetings, Party officials extolling the virtues of cooperation and hard work, and scenes of the evacuees being warmly embraced by their hosts in their new ("temporary") communities.[45] But what was really extraordinary about this film was a sequence in which the film crew was aboard a helicopter circling, not very high, above the smoldering remains of the reactor building. The voiceover, dubbed into English, was saying something about "black and white, the color of disaster." But what we *see* on the surface of the film itself are millions of tiny pops and scratches. The filmmakers explain that they had initially assumed they had used defective film stock. It was only later that they discovered that the problem with the film had nothing to do with the film itself. What they discovered was that the surface distortions were real field artifacts, and not defective film or processing. What was captured on the film was a record of the impacts of decay particles as they passed through the body of the camera. The irradiated film had captured a trace of the real, in this case a very striking pointillism of the real—discovered only after the fact, only retroactively. There is simply no correspondence of the film and its heroic worker narrative spin with the brute irruption of the real that is captured, incidentally, as its paradoxical urtext.

Žižek puts it this way:

> The paradox of the Lacanian Real, then, is that it is an entity which, although it does not exist (in the sense of "really existing," taking place in reality), has a series of properties—it exercises a certain structural causality, it can produce a series of effects in the symbolic reality of subjects.[46]

However, it is not as though the real is simply a raw material from which and upon which the symbolic makes a world. This may be partly true for the neonate (the confused, solipsistic, empiricist),[47] but it is the "real after the letter"—that is, the real that shows up in language as paradox and aporia—that concerns us here. In other words it may be helpful (after Bruce Fink) to imagine two different levels of the real: "(1) a real before the letter, that is a presymbolic real, which, in the final analysis, is but our own hypothesis (R_1), and (2) a real after the letter which is characterized by impasses and impossibilities due to the relations among the elements of the symbolic order itself (R_2), that is, which is generated by the symbolic."[48]

Sounding very Lacanian (which is odd), here is Peirce speaking out on threat in a way that anticipates Lacan's distinction between the real and the symbolic:

> "But what," some listener, not you, dear Reader, may say, "are we not to occupy ourselves at all with earthquakes, droughts, and pestilence?" To which I reply, if those earthquakes, droughts, and pestilences are subject to laws, those laws being of the nature of signs, then, no doubt being signs of those laws they are thereby made worthy of human attention; but if they be mere arbitrary brute interruptions of our course of life, let us wrap our cloaks about us, and endure them as we may; for they cannot injure us, though they may strike us down.[49]

This is most certainly not a kind of Wittgensteinian *What we cannot speak about we must pass over in silence.* Rather, it is *If it's not a sign, if it doesn't signify, then we shall not speak of it,* for only signs are to be the proper objects of our attention. This is exactly the problem. The interruptions are not mere. Something's status as a sign is surely secondary to its capacity to injure. They absolutely can strike us down, but they can also injure us—and the precise task of ecological thought is to figure out what this needs to mean for us. Which is to say that we must not just wrap our cloaks around us and hope for the best—things must also be made sense of.

Death

Question: And what does Death need time for?

Control: Death needs time for what it kills to grow in.

William S. Burroughs, *Dead City Radio*

From this perspective, to think of threat or its relation to the real, there is a constitutive derailment that must (somehow) be accepted. Threat of an ecological sort issues from the real, and the symbolic, as such, is incapable of jumping, suturing, or even fully apprehending that gap. This is key. Such threats as the "unrepresentability" of radiation—entirely chimerical "objects"—are manifest in a perfect indifference to our or any modes of symbolization. This, we could say, is the being of ecological threat; it presents itself, unrepresentably, as the threat of a death. Not just the death of the body, but something else. "Death insofar as it is regarded as the point at which the very cycles of the transformation of nature are annihilated."[50] Referring to de Sade *(System of Pope Pius VI)*, Lacan describes this second death: "Murder takes only the first life of the individual whom we strike down; we should also seek to take his second life, if we are to be even more useful to nature. For nature wants annihilation; it is beyond our capacity to achieve the scale of destruction it desires."

The first death is organic; the conjugation of "to die." I will die and you too will die, and so on. It is the death of the biological body. But the second death is different than this. The second death robs death from death. The second death annihilates the very cycles of life and death. For de Sade this would ultimately free nature from its own laws. But for us it points to the very kernel of ecological threat: the annihilation of the cycles of life and death and with it the symbolic universe within which it is staged.[51]

Chernobyl, then, represents a point at which the "open wound of the world" erupts, shaking the very ground of being.[52] For Žižek this recognition, this point at which the gap becomes visible, is only a point of re-cognition, that is, the gap was already there. The gap was already there. Žižek places it as the "unrepresentable point where the very foundation of our world seems to dissolve itself, there the subject has to recognize the kernel of its most intimate being."[53] And this is the constitutive nature of the gap, at least it is from a Lacanian perspective. On this view, ecological threats are thus a symptom of a prior disconnection.

The pathologies of response cannot find a resolution through tending to the needs of a nature-out-of-balance. This is simply not the problem to begin with. The presumption that nature knows—regardless which of the varied forms this may take—is simply and irreducibly a transference—and, it goes without saying, that this is in no way to diminish its significance. To speak then of the "adequacy" of response, with its resonance of Winnicottian good-enough, is not necessarily normative in a negative sense. It may be to rid discourse of its appeal to an imaginary equilibrium of a good-nature. For Žižek it is a call to "renounce the very idea of a 'natural balance' supposedly upset by the intervention of man as 'nature sick unto death.'"[54] Yet, whereas the notion of a second death animates some of the meaning of threat as I would like to develop it, an insistence on a constitutive gap or wound as the *condition humaine* fails to mark modern ecological threat as unique. And on this account, Chernobyl, and the manner of threat it poses, becomes only a figure; a figure in the sense that for Žižek, Hitchcock's films become figures for Lacanian concepts (this would be the threat as device, as pretext, as MacGuffin). To dwell upon the prior disconnection is to lose sight of the urgency and uniqueness of ecological threat.[55]

Mind the Gap

This Lacanian concern with the real and ecological threat is not exactly the invention of Žižek. Lacan engaged with the cultural implications of the real of nuclear threat in his 1959–60 seminar *The Ethics of Psychoanalysis*. Lacan's claim was that the contemporary condition is set apart, historically, as a function of the destructive power of the bomb. Hardly a remarkable claim but nonetheless. "I don't want to indulge in overdramatization," he said. Continuing:

> All ages had thought they had reached the most extreme point of vision in a confrontation with something terminal, some extra-worldly force that threatened the world. But our world and society now brings news of the shadow of a certain incredible, absolute weapon that is waved in our faces in a way that is indeed worthy of the muses. Don't imagine that the end will occur tomorrow; even in Leibniz's time, people believed in less specific terms that the end of the world was at hand. Nevertheless, that weapon suspended over our heads which is one hundred thousand times more destructive than that which was already hundreds of thousands of times more destructive than those which came before—just imagine that rushing toward us on a rocket from outer

space. It's not something I invented, since we are bombarded everyday with news of a weapon that threatens the planet itself as a habitat for mankind.[56]

So then what might we do instead? What other principles might inform response to threat? Žižek writes that we must come to a position that

fully assumes this gap as something that defines our very *condition humaine*, without endeavoring to suspend it through fetishistic disavowal, to keep it concealed through obsessive activity, or to reduce the gap between the real and the symbolic by projecting a (symbolic) message into the real.[57]

The threat of the real is a general imperiled condition—but it is one that also implies an ethico-ecological critique. The ethical basis of the real—or our relationship to it—is an often overlooked aspect of Lacan's thought. Or at least this can tend to be the case outside of psychoanalysis proper. It is here that Lacanian thought strays from a model of depth to something more approximating a practical ontology. To speak of the real one must of necessity speak of the real's relationship to moral activity, and to ethics as such:

The moral law, the moral command, the presence of the moral agency in our activity, insofar as it is structured by the symbolic, is that through which the real is actualized—the real as such, the weight of the real.[58]

In a way this is nicely preworked for our purposes. It offers a way of thinking about the gap as an ethical site. Symbolic activity around the real of threats—which runs the full spectrum from the WIPP's desert burial to political debate about global warming, the genetic manipulation of food, stem cell research, contaminated water, molecular farming, and the air you are breathing right now—is for this reason an ethical engagement. Identifying, naming, understanding, classifying, prioritizing, all of this involves invention. What do we make of a Chernobyl, a global warming, a contaminated water supply? The changes these things introduce into a world of subjects do not spring forth fully formed. That is, how we come to accommodate the unwelcome intrusions, displacements, and upsets posed by ecological threats—the weight of the real—is not only instrumental but ethical as well. And this is precisely why the real is of interest. Ethical action is "grafted onto the real . . . introduces something new into the real."[59]

So far so good. Mind the gap: this is good advice. But the real may also unwittingly foster a kind of fatalism with respect to our own relation(s) to ecological threats, as well as the limits of thinking about the future in relation to them. I mean this in the sense that if the real is the event that always arrives largely unannounced, with no warning, then the ethical reflection that Lacan has said lies at the core of analytic practice with respect to the real is called to an encounter only after, only after it is too late. It is the debt as effect that arrives first, making the call to ethical reflection "too late"—the real is traumatic, and the repetition that ensues does so without the possibility of paying the debt now owed, the wound, the gap, that is opened by the intrusion of the real. It is the retroactive discovery of the real's irruption into life that is the ethical call, and it is "too" late that the call is heard. It may be too late for many things. The Alliance of Small Island States was making exactly this claim in Kyoto. It will be too late for us, they said—meaning it was already too late. Their claim was not about standing, but about existing. And it was too late for Chernobyl when the expected arrived unexpectedly and too late as well for those things that have not entered public awareness or for those that did but have slipped away. The wall of mud and rock and debris that smashed into the Vargas area north of Caracas, Venezuela (December 1999), displacing the already displaced, flashed up only briefly as a concern. But when the Payatas dump outside of metro Manila (home for a third of Manila's daily domestic waste—four million or so households—and as a result, for some twenty thousand scavengers) collapsed in June 2000, the thousand plus who were buried alive hardly registered at all. But this is about something else, perhaps about what it is that moves one to care. The accident, though, stands apart from this. The accident: it collapses the too soon and the too late, leaving us, in its early (and thus anticipated) arrival, too late to respond?

The chimerical "objects" of threat are both an instance of the derailment, the gap, and a point at which the call may be heard. Or not. This is the problem. The call is heard, or not. And the intrusion, such as it is, is constituted as a surprise, a shock, an answer, or nothing at all. Or a "problem."[60] In an analytic situation the problem is exactly this "or not." The sound of the call itself is the problem. From an analytic point of view, one works toward the symbolic establishment of an accounting, a "draining away" of the real, as J.-A. Miller put it. But we are not in analysis here. Threat may be the royal road to the real, but it is necessary to establish its implications *for us*. On one hand, the movements of threat, nuclear or otherwise, are difficult to recognize, and even when we do, may seem insufficiently traumatic to move us to action. And on the other, the routineness of accident and disaster have fostered a situation of ontological numbing to the suffering of others. Both too little and too much. And in any case, the complexity of social and symbolic life arrays its defenses to keep threat on the side of the symbolic—threat as risk (heuristics), threat as side effect (epidemiology), threat as recompense (nature bites back) or sign (nature's commentary), or threat as unintended consequence of whatever (shit happens). The real is a problem. And it leads to another problem, to trauma.

Trauma

If the real's intrusions could be rapidly, reasonably, and adequately accommodated into one's symbolic universe then there would be no problem. This is hardly the case though—if only for the reason that we have no idea what rapidly, reasonably, and adequately might mean in this context. Is there an irreducible traumatic quality to the persistence and fact of ecological threats? What does it mean to be threatened, to be under threat? It is not the same thing as fright, or fear, or anxiety of some sort. These states are not equivalent, or synonymous. "Anxiety," Freud explained,

> describes a particular state of expecting the danger or preparing for it, even though it may be an unknown one. "Fear" requires a definite object of which to be afraid. "Fright," however, is the name we give to the state a person gets into when he has run into danger without being prepared for it; it emphasizes the factor of surprise.[61]

Nor do these things capture the specificity of ecological threats.

Threat cannot be prepared for, it is not a particular "object" to which a fear may be attached, nor is it simply a shock or surprise that frightens. The kind of injury that threat produces is difficult to see; it is as though we are wounded though cannot express that fact, nor how it came to be.

We've considered some of the mechanisms of defense that occlude our vision in both directions (that is, of the threats and our relation to them). Perhaps the only explanatory feature we can offer at this point—although I do so with a certain trepidation—is trauma. There has been such a great deal of renewed interest in trauma theory of late, that by being everywhere in a near pan-traumaticism, it threatens (so to speak) to cancel itself out. We do not need to become involved in the mimetic versus antimimetic debates that have been laid out with such clarity and detail by Ruth Leys.[62] That said, what is of significance here is the movement of trauma away from the limited and bounded sense of a blow or injury sustained to the body and toward a sense of trauma that encompasses the individual and, ultimately, the social.

As Freud put it:

We describe as "traumatic" any excitations from the outside which are power-ful enough to break through the protective shield. It seems to me that this concept of trauma necessarily implies . . . a breach in an otherwise efficacious barrier against stimuli. Such an event as an external trauma is bound to pro-voke a disturbance on a large scale of the functioning of the organism's energy and to set in motion every possible defensive measure.[63]

The majority of recent trauma work is engaged explicitly or otherwise with a Freudian reading of trauma. In *Beyond the Pleasure Principle*, Freud reaches a point where he must deal directly with the seemingly paradoxical force of trauma.[64] The binary model he had established—pleasure principle/reality prin-ciple—became exceedingly difficult to support in the face of the specific pathologies he witnessed in war neuroses and survivors of war. Traumatic neurosis, at least inso-far as it interacts within the pleasure/unpleasure economy, was more than a conun-drum. Whether we view this moment in Freud's work as the insertion of an epicycle onto an unwieldy theoretical apparatus—the death drive is often said to be the point at which Freud slips most directly into speculative anthropology—or the point at which he most directly grapples with the *human condition*—he clearly showed how trauma is an "event" unlike any other.

Trauma is marked by two necessary features. The first is that the traumatic event represents an experience that exceeds one's capacity to experience and to understand. Accordingly, it has a quality of a paradoxical experience. It is to *have been there*, yet to be unable to integrate the experience into one's biography, into one's practicable universe. An experience, as Dominick LaCapra put it, that is not fully owned—in the sense that although one has had it, it is not something one possesses.

But it is more than this. Trauma is something that effectively hap-pens after it happens. It is experienced as the effect preceding—indeed eclipsing—the cause. The unrepresentability and unassimilability of the traumatic event when it occurs sets up a hole in the subject's symbolic universe; it becomes a place where the symbolic falters, and yet something about the event is preserved (we will not comment here upon the literalness of that preservation). Such experience is to suffer the effect of a causeless cause. Trauma is the nonplace that stands as the location of limit events, a foreign and strange place in the subject. It is like a gap that is marked by the absence of an experience that was undergone but not registered and is given only in its evidence by its effects—flashbacks, recollections, dreams, hallucinations, etc. That is, symptoms.

Trauma persists, then, somewhere between an event and the im-possibility of that event's symbolization. Accordingly, from a Lacanian perspective, trauma is a relay between the real and the symbolic. Yet the mode of its connection is obscure. On the one hand, Lacan explained that the real is always a kind of encounter that is missed, *essentially* missed. Yet on the other, this encounter is somehow preserved and marked with such an insistence that it—or scenes *of* it—are subject to repetition. This is the haunting quality of trauma: it is to be subjected to repetitions of something for which there is no original. "It is through its 'repetition,' through its echoes within the signifying structure, that the cause retroactively becomes what it always-already was."[65] Writes Žižek:

> The cause qua real intervenes where symbolic determination stumbles, misfires, that is, where a signifier falls out. For that reason, the cause qua real can never effectuate its causal power in a direct way, as such, but must always operate intermediately, in the guise of disturbances within the symbolic order.[66]

To further complicate matters, there is an inherent difficulty, because of the retroactive aspect of trauma, in determining and sorting out the traumatic memory in relation to the historic traumatic event. Some research into post-traumatic stress syndrome suggests that the experience of trauma disrupts declarative memory, but not the nondeclarative or implicit memory. From this point of view, while the intentional recall of the "traumatic event" is constrained, that part of memory responsible for emotive and affective responses and sensations related to past experience is not. In Freud's Little Hans analysis, he writes, "A thing which has not been understood inevitably reappears, like a ghost, it cannot rest until the mystery has been solved and the spell broken."[67] Jean Laplanche, commenting on a passage in Freud's "Project for a Scientific Psychology" (where Freud states that "we invariably find that a memory is repressed which has only become a trauma *after the event*,"[68]) writes:

> Here is the heart of the argument: we try to track down the trauma, but the traumatic memory was only secondarily traumatic: we never manage to fix the traumatic event historically. This fact might be illustrated by the image of a Heisenberg-like "relation of indeterminacy": in situating the trauma, one cannot appreciate its traumatic impact, and *vice versa*.[69]

Disaster: Outside, Inside, Both

The first derivation of trauma was from the physical to the psychical, from the outside to the inside.[70] The second derivation we could say is from the psychical sphere of the individual to the collective. (This is less a reversal—moving back to the outside—than it is to posit something either contiguous or analogous between the two spheres.) In any case, if the first move was difficult, the second is even more so. The resistance and inertia are considerable. To begin with, limit events or disasters are not characteristically thought of from a point of view of trauma. (At least not until quite recently—post-Vietnam—and even here it becomes a psychiatric not psychoanalytic category.)

There are exceptions to this, but for the most part there are deeply ingrained and habitual ways of thinking about disaster that stand in the way of consideration of social trauma(s). The enigmatic quality of disaster has kept attention on questions concerning how one defines the disastrous event—that is, what is a disaster?[71] That these roughly correspond to successive disaster or large-scale hazard research paradigms makes them all the more exemplary as expressions of how the object of threat and its actualizations in accident and disaster have been identified, constructed, and legitimated.[72]

The first is the thought of disaster as corresponding to an attack visited upon humans from a hostile outside agency: a Hobbesian Nature—nasty, harsh, brutish, and so on. Here humans are the recipients of an unbidden and unexpected visit from a capricious nature. They (we) come under siege. In other words, it is war by other means (or perhaps the very first means[73]). This is of course by far the most prevalent and familiar mode of both disaster conceptualization and response. One needs only to consider the equation between "humanitarian intervention" on one hand and disaster aid and military organization and effort on the other. In the face of a hostile nature, science (technomilitary) alone is capable of a response.

A companion piece to the nature-as-aggressor model would be disaster conceived instead as an expression of human vulnerability. Here it is not about the ultimate power of nature but rather the vulnerability (understood socially, biologically, technically) of human society, of human communities. This is merely the distorted opposite of the first view, and like many oppositions, it manages to preserve the explanatory force of the first view within its apparently different thrust. Disaster here expresses human vulnerability (an "empirical falsification of human action"[74]), but still there is a Nature that is poised to wreak havoc upon the similarly defenseless humans. On this account what disasters do is reveal the contradictions of inner social logic. Thus the Concorde didn't crash. What did happen, though, was that the contradictions implicit in the social, historical, and political organization of the French and international airline industries became briefly and tragically exposed (a manner of reasoning that on the face of it—e.g., try it with September 11—seems far more rhetorical than sociological).

The third thought of disaster—although not prevalent anywhere but within certain reaches of disaster sociology—but still very much related to the first two (a compromise-formation we could say) both does and does not preserve the other two. Here it is uncertainty. Uncertainty is the absence of explanation reducible to either nature as hazard or the social as vulnerable. It allows for complex causality and seeks explanation of the disaster as an interruption in the social fabric of meaning. The disaster here can be understood as a crisis in information and, thus, communication. Accordingly the only useful response can be the reduplication of technological and scientific effort because such things alone are "able to probe and domesticate these further reaches of environmental and social 'wildness.'"[75]

In each case the social is subject to this something that is disaster, either as an outside force (requiring a social-technical-military mobilization), or as an expression of a vulnerable and insufficiently armored *inside* (likewise requiring the reduplication of social "security" measures—*it's for your own good!*), or as a field of complexity (requiring above all increased technological incursion into the chaotic [im]balance of the social/natural).

Communities of Disaster

In the overall arc of disaster research there has not been a great deal of work that attempts to link these events as trauma. Rather, disaster literature has tended to emphasize disasters foremost as being events; events moreover that result in an upsurge of community togetherness, caring, selflessness, cooperation, and so on. It is as though the instinct for survival, and in the aftermath, the feeling (guilt) of *having been spared*, results in the suspension of the pettiness of the everyday and a surge of humanitarian goodness.

In a frequently cited study on the psychology of disaster that seems to have been important in the promotion of this particular idea, Mary Wolfenstein wrote:

> Following a disaster there is apt to be a great upsurge of good will and helpfulness among the survivors and on the part of outsiders who come to their aid.... Those who have undergone the impact of a disaster have in that moment concentrated their emotional energies on themselves. Afterwards there is a compensatory expansion of feelings towards others, partly motivated by the guilt of not having cared what might happen to them when one's own life was in danger. In the moment of impact...the victim is apt to have an illusion that he alone is affected and to suffer painful feelings of being abandoned by others and by fate. The discovery that one did not suffer alone and

the sight of friendly hands held out to help one are all the more prized against this background of loneliness. Also, having been chastened by the punishment of disaster, one is eager to be exceptionally good to make up for past derelictions and to ward off further retribution.[76]

The point she is working toward in this passage is that there is a postdisaster utopia in which survivors—through guilt at having survived or sorrow for those who did not—rise up and overcome. These are simply the good Christian overtones of surmounting adversity, putting aside one's petty interests in the interest of the greater good, etc. Variously termed "city of comrades," "democracy of distress," "community of sufferers," "altruistic community," there has been a tendency to foreground disaster as a prelude to rebirth.

Nonetheless, there have been some breaks with the orthodoxy of this thought. One of the most interesting and provocative challenges to the sway of this particular tradition in sociological and psychological disaster theory—and, as we will see, a singularly promising instance of contemporary trauma theory—has come from the sociologist Kai Erikson.

In 1973 Erikson was involved as a consultant (for litigants) in the aftermath of a disaster that occurred in the Appalachian community of Buffalo Creek, West Virginia. In this disaster, a coal mine tailing-pond embankment burst, sending a wall—some million and a half gallons—of mud and debris down a very narrow valley that contained the homes and histories of the five thousand residents. In less than five minutes, one hundred thirty-two people were killed, and four thousand were left homeless. Erikson's book, which chronicles his time in the remains of the Buffalo Creek community, is a fascinating piece of disaster fieldwork.

Yet, as one works through Erikson's text on Buffalo Creek, one gets the quite remarkable sense that here is a sociologist who—in the course of his investigation—essentially runs out of reasons to support the received orthodoxy of disaster research. Indeed, the situation at Buffalo Creek could not be made to conform to any of the standard assumptions about the wake of disaster. Not only the individuals but the community was profoundly damaged in ways that he was not given to expect. The survivors did not see their lives having been spared as a moral accomplishment: the spirit of community-building was nowhere in sight.[77]

In the conclusion to *Everything in Its Path*, Erikson suggests a slight but important shift in thinking that he sees as necessary to the development of the idea of collective or community trauma. Specifically, he suggests that rather than seeing trauma as an *effect* of some manner of injury—in other words, of finding in trauma a causally induced condition in the wake of the "disaster"—that we reverse the procedure. This would mean that the important criteria become the traumatic reaction and not prevailing definitions of "disasters."

> In the first place, we would be required to include events that have the capacity to induce trauma but that do not have the quality of suddenness or explosiveness normally associated with the term. For example, people who are shifted from one location to another as the result of war or some other emergency. . . . And one might add here that thousands of American Indians, confined to reservations for the better part of a century, continue to show effects of traumatization. Our list might also have to include such slow developing but nonetheless devastating events as plague, famine, spoilage of natural resources.[78]

By opening up the concept of disaster so that it, too, may be retroactively constituted and so that we may come to form a concept of the slow-motion catastrophe, trauma gains a kind of mobility and diagnostic scope not traditionally accorded it. Disaster is a paradoxical event too. By changing what can count as a disaster, one approaches the notion that

> *chronic conditions* as well as *acute events* can induce trauma, and this, too, belongs in our calculations. A chronic disaster is one that gathers force slowly and insidiously, creeping around one's defenses rather than smashing through them. The person is unable to mobilize his normal defenses against the threat, sometimes because he has elected consciously or unconsciously to ignore it, and sometimes because he cannot do anything to avoid it in any case.[79]

And in the concluding passage to this work, he offers:

> I have suggested that human reactions to the age we are entering are likely to include a sense of cultural disorientation, a feeling of powerlessness, a dulled apathy and a generalized fear about the universe. These, of course, are among the classic symptoms of trauma, and it may well be that historians of the future will look back on this period and conclude that the traumatic neuroses were its true clinical signature.[80]

For our purposes here, the key insight that Erikson makes is that the disaster is not solely about an event of some kind, for the disaster may come without an event. Rather, it is: if there are those who are traumatized, there has been a disaster. (Note that I do not say "if there are victims there has been a disaster."

This is because not only victims may be traumatized, that is, perpetrators may be as well, but it hardly makes sense to also make of them victims.)

Freud also had suspicions about the well-being of communities. In 1929 in the conclusion of "Civilization and Its Discontents," he made the link between individual development, vis-à-vis neurosis, and the development of systems of civilization.

> If the development of civilization has such far-reaching similarity to the development of the individual and if it employs the same methods, may we not be justified in reaching the diagnosis that, under the influence of cultural urges, some civilizations, or some epochs of civilization—*possibly the whole of mankind*—have become "neurotic"?[81]

If we are to say that trauma is an unmediated and excessive event that overwhelms by its very excessiveness, there is no reason why we should not come to speak of groups rather than individuals. Some writers, for example Robert J. Lifton and Robert Kaplan, have been developing profiles of individuals and communities in the wake of disasters of various sorts. Kaplan, in *The Ends of the Earth*, has taken travel writing to its limit in the form of a disaster travelogue—truly an *atlas calamitas*—of "Third World" social, political, ecological, and historical breakdown.[82] Lifton's work spans decades and has been concerned not only with extreme traumas but with the manner in which the encounter (therapeutic or otherwise) with survivors creates a secondary traumatic subject position of the *proxy survivor*.[83]

Holocaust writing contains the most concerted attempt to understand the social (and transgenerational) dimensions of trauma. From, for example, Saul Friedlander, Shoshana Felman, Art Spiegelman, Primo Levi, Claude Lanzmann, and Jean-François Lyotard, there have been highly diverse strategies used to confront the collective and historical condition of Jews (in Europe and elsewhere) in the wake of the Holocaust. And without seeming to use the Holocaust as *an example*, one must nonetheless say that there are others. There are countless others. The recent short list would include Hiroshima, Nagasaki, Dresden, Vietnam, Cambodia, Bosnia, Serbia, Kosovo, Uganda, Rwanda, Zaire, Guatemala, and Chile. What unites this disparate geography is disaster. Yet these examples constitute some of the most powerful features of our time. Too powerful, really, in the sense that one easily becomes caught up in the profound and overpowering tragedy and suffering of these places, peoples, and events. The *atlas calamitas* of the past century (and one needn't stop there) is just about as large as we allow it to be. And just about every imaginable psychological, social, and political defense mechanism has been employed against this understanding. In a sense this supports what I'm trying to get at here; the magnitude of these events, in threatening one's capacity to comprehend, summons instead protective mechanisms of defense.

It's a dosage problem to a certain extent; the dose needs to be titrated. As LaCapra says, we need to find a homeopathic dose—a homeopathic repetition of events—in order to work with such things.

> One may maintain that anyone severely traumatized cannot fully transcend trauma but must to some extent act it out or relive it. Moreover, one may insist that any attentive secondary witness to, or acceptable account of, traumatic experiences must in some significant way be marked by trauma or allow trauma to register in its own procedures. This is a crucial reason why certain conventional, harmonizing histories or works of art may indeed be unacceptable. But one may differ in how one believes trauma should be addressed in life, in history, and in art. Freud argued that the perhaps inevitable tendency to act out the past by reliving it compulsively should be countered by the effort to work it through in a manner that would, to some viable extent, convert the past into memory and provide a measure of responsible control over one's behavior with respect to it and to the current demands of life. For example, the isolation and despair of melancholy and depression, bound up with the compulsively repeated reliving of trauma, may be engaged and to some extent countered by mourning in which there is a reinvestment in life, as some critical distance is achieved on the past and the lost other is no longer an object of unmediated identification.[84]

Our task, if we are to become better equipped to deal with the paradox of ecological threat, involves several things: the first is to know that the impulses toward an objective picture of threats is necessarily fraught with projections and displacements. We need to know when our motives and our behavior are determined through various forms of acting-out, through unthinking repetitions, and disavowals. From a Freudian point of view, this is a call for a working-through—that is, a modified repetition that is supplemented by interpretive attempts to understand the repetitive formations (and the resistances that found them).[85]

It is of critical importance to explore further the implications of the question about the collective trauma posed by ecological threat. LaCapra suggests that it would need to become the site of a *traumatic transvaluation*.[86] A transvaluation that would in a sense capture what Ewald sought in his example of the survivalists. But rather than communities founded solely upon survival, we may become free to imagine new identities, strategically claimed to account for the presence of threat in a way that neither bypasses it through psychological mechanisms of defense nor consigns us to an infinity of traumatic repetitions. Obviously one must be careful here. It is not a matter of *equating* threat and trauma—logically speaking this would leave the middle (the real) undistributed.

In one sense, nuclear and ecological threats mimic the structure of trauma. That is, because we focus on the event (on threat's actualization, on the disaster) ecological threat can itself be retroactively constituted and, thus, gain a kind of uncanny aura, leading to a condition of paranoia of toxicity and contamination. (The recent and possibly ongoing bovine spongiform encephalopathy meat apocalypse in Europe, the United Kingdom, Canada, and the United States provides abundant evidence of this and, more recently, anthrax in the United States and elsewhere.) And in another sense such threats are precisely traumatic for the reasons noted above. Culturally and socially there has been virtually no opportunity to work through the advent of nuclear threat. The modern history of the nuclear was inaugurated by a stunning mass murder, moved into the high anxiety of the Cold War, and gained attention briefly as an environmental issue and then an energy issue, and now we wish to commence with its ending with a quiet burial (even as the electrical utilities are once again painting a picture of a clean and safe nuclear future).

On Possibility and the Virtual

We need to return again to the real. Lacan's gradual elaboration of the real as constituted by a retroactivity, a backward look, seeks an understanding of that which has already taken place. The real, as that which intrudes—and as that which remains necessarily a conjecture—is always posed in terms that require an accounting: What has happened? Why do I enjoy my symptoms and want, above all, their efficacy restored?[87] Why are you always so pessimistic anyway? Why do I feel pain? The unassimilable "nature" of the real, its retroactivity, and its numerical quality of (ac)counting, point to its didactic—if paradoxical—function: knowledge of the gap. The real itself is the "surprising fact"; the result of an abductive leap that seeks to account for a state of affairs through an elaboration of what has happened without our knowledge of it. In this sense, our knowledge of the real is always a kind of damage control. Our knowledge seeks explanation of the *already*. The real, as such, is within a past tense, and for this reason difficult to see as performative or productive. (Or at least productive of something other than ciphers of its own intrusions and pathologies.)

(And the real itself is simply too large. For the intrusions of the real include many things that are not, in a significant fashion, historical events. In other words, if the real is traumatic, it is productive of very different sorts of trauma. There is a difference between oedipal conflict, the entrance into language [R_1] and incest [R_2], on one hand and an earthquake on the other. Seeing in both the same sense of trauma is to allow for a constitutive sense of trauma [trauma as the paradoxical ground] making everyone already a victim or survivor, which comes to the same.)

If the real, this Lacanian real, leads us to an acquiescence, if it leads us to a point where all we can hope to accomplish is to figure out what has taken place by reading the tectonic shifts and sifting through the debris of our symbolic world, then why bother with it? The answer is simple: the real is a challenge to rethink ecological threat.

The real is the do-nothing hypothesis, the null hypothesis of threat. (Perhaps this is why it is so easy to demonstrate the concept of the real by using examples of disaster.) Even though it is wrapped up in psychoanalytic garb, a concept of the real is already contained in (or implied by) the conventional response to threat and disaster. How else to account for the perpetual and habitual astonishment when bad things happen? The real is there in the symptoms.

So in light of this, and equally, in light of the difficult pathologies of response, we need to drain a lot more of it away. This means simply that the challenge of the real is not to make it signify, it is not to render it coherent, it is not to reduce it, to displace it, to contain it, to rename it, to administrate it, or to otherwise capture it into preexisting categories or concepts of risk. The challenge of the real of ecological threats is precisely to discover a mediator that will allow something new to be said, that will perhaps allow a qualitatively new manner of thought and action to inform a time (ours, for instance) in which the productive capacity of threats seem to outstrip any reasonable capacity for reflective (affective) response.

So perhaps here we need to be more precise about what the concept of the real can continue to provide. It is difficult to imagine dispensing with the concept entirely; it follows language like a phantom. The suddenness of overwhelming events (the gas leak, the earthquake) and the horrifying retroactivity of the toxic slow motion event (Grassy Narrows, Love Canal) seem so clearly to be irruptions of a senseless *outside*. To the extent this is true, it matters little whether we conceptualize the tropical storm as a hostile incursion of nature, a glimpse of human haplessness, or a chaotic event. It may make no difference to the body count. But from another point of view—the point of view of threat—it makes all the difference how we think about such events. The Lacanian real leaves us few options. It describes a manner of intrusion of events. It highlights the (paradoxical) retroactive activity of naming and understanding, of trying to make sense of the senseless, and it highlights this as an ethical activity. But aside from the fact that we are constantly naming the real, that we are constantly symbolizing that which resists symbolization, making fissures in that which has no fissure, the real's chief defect is that until it intrudes it is not there. (Until there is a disastrous event, it's business as usual.) Even though the real is a radical tool for understanding the limits of language, for seeing the working of the symbolic, and for theorizing the hard parameters of trauma, the real is simply not radical enough. To operate with the real as a theoretical horizon is simply not enough. To persist within its gravity one has already agreed to a lot, tacitly at a minimum. One is already locked into a model of depth—if nothing else the real presupposes an unconscious, a vertical model of self-concealment. One is already locked into a model of retroactivity and belatedness—the real (together with its pathology-by-proxy, trauma) is never present to (an) understanding, yet it always reveals itself (eventually) as an image of what has already taken place. Critically speaking then, the real functions as a kind of deconstructive figure that takes us a certain distance, but only on the condition that we leave it aside when it ceases to function.

The real has no positivity, all it does is leak, all it can do is make trouble elsewhere. Yet when we think of this from the point of view of threat, this is not our experience at all. Threat precisely has a positivity, it just has no actuality. In itself it presents nothing for measurement and must come down to a social fact (which is exactly why one must have recourse to concepts pertaining to response—mechanisms of defense, and so on). So the suggestion here is that a shift must be made. Traditionally the natural disaster had been the model for understanding the unexpected event. When disaster becomes an object of knowledge itself, we end up with disciplines of knowledge that attempt to transform and domesticate disaster by making it a function of probability. Threat can only then be an expression (if at all) of a prior probability, understandable as the key variable in the calculus of the disaster. If we reverse this procedure, however, and look at the disaster secondarily, as a production, through the optic of threat understood in its positivity, the picture begins to change. How can threat become the model, in other words, and disaster an expression of it?

The virtual is just the shifter or mediator we need to think of threat in its positivity and to transform its disastrous incursions into something other than terrifying ciphers. The virtue of the virtual is that it allows us to see that threats themselves produce transformations, that they are not merely prefatory to an event (ineffectual, but ambiguously prognostic), and to see that these transformations are difficult to recognize through an optic of risk and possibility.

To think of threat as virtual (and disaster as one of its actualizations) may come to make a real difference in how we might think about ecology, about disaster, about technology, and the future. (As to the virtual itself, I have stolen this concept, and I use it in a way that is partially faithful to its author, and at same time treacherous—and therefore perhaps even more faithful.)

For Deleuze, the virtual "has a reality proper to it, but that does not merge with any actual reality, any present or past actuality."[88] In *Difference and Repetition*, Deleuze's refrain of the virtual is this:

> The virtual is opposed not to the real but to the actual. *The virtual is fully real insofar as it is virtual.* Exactly what Proust said of states of resonance must be said of the virtual: "Real without being actual, ideal without being abstract"; and symbolic without being fictional. Indeed, the virtual must be defined as strictly a part of the real object—as though the object had one part of itself in the virtual into which it plunged as though into an objective dimension. . . . The reality of the virtual is structure.[89]

Deleuze here is outlining (via the philosophy of Bergson) a philosophy of ontology conceived as a creative force of Becoming rather than the static state of Being(s). And, as always, when Deleuze is teaching a philosophy lesson, he is also offering practical advice for thinking and living. The traditional opposition, postulated by philosophy and theology, and internalized by culture (an image of thought), is between the *possible* and the *real* (although *not* the Lacanian real). On one hand, that which *is*, and the other, that which could possibly *come to be*.

Logically, the possible must precede the real (for it contains the prototype of the real thing prior to its becoming real). Conceptually though, the only difference between the two is that one exists and the other does not. Think of a headache. The headache you actually have is different from the one you could have possibly had an hour earlier only in virtue of the fact that you now have it. It is different only in the sense that what you now have is actual, and what you might have had before was only a possibility. To put it differently, all you have to do is add the single ingredient of *existence* to the possible headache and voila, you have the actual thing.

In this sense exactly, in thinking about how it is that things come to be such as they are, there is a kind of "retrograde movement of the true" according to which an image of existent things is projected backward into a prior possibility; a nonbeing that existed prior to being.[90] So this leads to the false conclusion that prior to being the case, everything exists as a possibility.[91]

> That is why it is difficult to understand what existence adds to the concept when all it does is double like with like. Such is the defect of the possible: a defect which serves to condemn it as produced after the fact, as retroactively fabricated in the image of what resembles it.[92]

Thus, the real is already given in the possible (because it was already there in a sense, waiting *qua* possibility); it simply "has existence or reality added to it...from the point of view of the concept, there is no difference between the possible and the real."[93]

From this point of view, the process of actualization—that is, the movement from the possible to real—is subject to or governed by two rules or constraints: limitation and resemblance.[94]

In the first instance, the possible is taken to be an image of the real that—through a process of limitation—either does or does not get realized. This is simply to say that there are more possibilities than there are real things (which is also to say that the realm of the possible is larger than the realm of the real). Therefore, in the movement from the possible to the real there is a kind of funneling or limitation. The possible is thus a kind of preformed proto-reality or "pseudo-actuality."[95] And the real is consigned to the realm of resemblance. Existence—*being real*—is therefore just a doubling with what was already there as the possible. After all, "what difference can there be between the existent and the nonexistent if the nonexistent is already possible, already included in the concept and having all the characteristics that the concept confers upon it as possibility?"[96] No creation, just identity and doubling.[97] "Everything is already *completely given*: all of the real in the image, in the pseudo-actuality of the possible."[98]

The second constraint that Deleuze foregrounds is resemblance. Since apart from existence there is no difference between the possible and real, then the real and the possible already resemble each other. The real must necessarily resemble what was already given as possible.

Threat works its way into this precisely in the sense that as virtuality it cannot be said to exist in the sense that something actual exists. Nor does it exist as an image, or prototype, of an event or occurrence that might become realized. Rather, threat *subsists* (Brian Massumi's term) as virtual *and* real.[99] This movement of the threat of the nuclear concerns us in two ways. First, in the movement from virtual to actual (the *event*, the accident) there is an actualization in which what was virtual becomes swept up into a specific social configuration. To paraphrase Deleuze, there is no total threat in which all possible kinds of threat are incarnated; rather there are specific instances in which certain elements of the entire virtual field of threat becomes actualized.[100] Yet we must also say that as threat becomes actualized, it becomes something else. And the "else" it becomes is dependent on the host of specific relations into which and through which it becomes actualized. This is exactly the sense in which it can be conceived of as a creative force; there is no simple mimetic or metaphorical correspondence between threat as virtual and the event. And second, as regards threat as virtuality, we have something else entirely. Real but not actual, threat has no existence of its own apart from its various actualizations. And its paradox—formulations of which we noted above—is that it can only be read in its effects. "Structures are unconscious, necessarily overlaid by their products or effects.... One can only read, find, retrieve the structures through these effects."[101]

To bring the virtual into the discussion about threat presents a number of interesting openings. First of all, we may observe that discourses of risk— the very idea of a science of risk—proceeds entirely from this frame of possible-real. Risk thought imagines that events are known as possibilities before they become real. This much is obvious. You either had a car accident or you didn't, and if you did, someone was either injured or not. So when risks become events, when they become real, they accord on one hand to a measure of probability and on the other to an image of the possibility of which they were but the literal representative.

The entire project of the WIPP, from its inception in the late 1950s to ideas about the stability of salt formations, development of human intrusion scenarios, projection of future social organizations, modeling of transportation accident scenarios—all of this presupposed the development of a set of possibilities of which some (and not others) will become real, and furthermore, those that do will perfectly resemble the possibilities they incarnate. And if they do not, when the unlikely accident happens it is written off as poor science, or outside the bounds of a risk scenario (for example, more than the credible accident), or human error (the final frontier of the technological alibi). Probabilities are like pictures of accidents, disasters, failure modes, and mishaps. As pictures they are likenesses of things, they are wishful iconic schemas that we imagine to be indexical—as though the probabilities really were connected with the events. In fact, the only indices are the events themselves; indices par excellence.

Risk fails to account for too much. With ecological threat the errors and omissions of risk have already, and long ago, become perilous. Projecting this manner of thinking into the deep future is a spectacular error—one would imagine that there is just too much *very real* evidence in the form of unlivable habitat, altered systems, damaged humans, and dead bodies, to justify such a staggering leap of faith (assuming that's what it is).

But in any case, we only need to empirically test the assertions of probability. Experientially, this is all constantly, relentlessly, and mercilessly subjected to falsification. Today's accidents, ecological accidents, technological accidents simply never follow this line. This is, in part, what makes them so profoundly frightening. And it points directly at a social paradox. Social institutions (technological, administrative, scientific) behave directly within a paradigm or philosophy of the possible. The rest of us, however, to the extent that the various perturbations register at all, really are left to *wrap our cloaks about us* as planes and buildings fall from the sky and hope for the best. (No wonder we're all a bit messed up.) Living under ecological threat and simultaneously living in a time that either refuses to allow threats to register as concerns or transforms them into something entirely different—this is not the mark of a culture that ought to be making decisions that organize the next ten thousand years. Much of the traumatic force of ecological threats—though obviously not all—is precisely because so much social comportment feigns otherwise. While ecological threats cannot be willed away, they can become *less* traumatic. Paul Virilio's suggestion that we (socially) work out what kind of accidents we can live with and then build the technologies those accidents presuppose is not far off the mark. It suggests a fundamental principle that underlies all of this: ecological threat demands a new analytics of the accident.

Virilio says, "The breakdown of the American nuclear plant at Three Mile Island calls into question the breakdown of war, nuclear deterrence, and thus in the long run peaceful coexistence itself: the intense publicity surrounding the event and the risks incurred by the people in the area *transforms the lifestyle question*."[102] And what he means by this transformation is that "*that kind of threat*" alters the psychological behavior of the society concerned.[103] And the transformations wrought are varied, entirely real, and not subject to an event (or they constitute an event in their own right). They cannot be written off as mere neurotic and paranoid preoccupations. The basic conditions of meaning have changed. A huge claim, by which I mean this: prior to ecological threats, to say "I worry about the future," no matter how deeply existential the motivation may be, cannot mean the same thing as *after* the acknowledgment of such threats. In the former case one may mean that one worries about the state of the future, and perhaps one's mark upon it, but one would have no reason to mean *whether* in fact the future will take place.[104] That is, one would have every reason to believe in some manner of continuity. Today, this is no longer the case. The second death, or ecological death, fosters a double blindness. We are blinded to the fact of ecological threats generally—from one side, the nuclear weapon and the trope of the mushroom cloud having decisively usurped the apocalyptic imagination, and from the other, until the accident actually happens, probability keeps things *uneventful*. But they blind us too to the circularity of the logic that tethers us to solutions in the form of further technoscientific incursions. If we think back to the magnitude of each of the variables in Diefenbaker's equation of nuclear war—that is, "us," "them," or by "mistake"—we see that it has been a shifting landscape. That is, the "mistake" (he called it miscalculation), tacked on in the heat of the Cold War as a preemptive alibi and token of human fallibility, now comes to organize the entire field (or nearly). In order to continue with nuclear practices (both weapons and energy) techniques of administration must become realigned toward the probability and time of the "accident."[105] But for the grace of a few seconds, or hours, or days, the incidents at Harrisburg and Chernobyl were prevented from realizing their full potential. Time becomes the key variable, and risk becomes the discourse that supports time's administration (including the time of the casualty not yet born). We see this raised to near infinite proportions in the case of the burial of nuclear waste. But we see it elsewhere as well. The Kyoto Protocol can be read as a temporal agreement that trades time against indulgences of disavowal. And in all of this, a society is transformed administratively and psychologically to account for the presence of threat. In this sense, a politics of (nuclear) war permutates into the politics of risk.

Outing Threat

Threat is everything not accomplished by the accident. This of course makes it no less real, nor does it privilege the event of the accident.

This is precisely the direction we need to travel in order to understand the workings of ecological threat. Strategies cannot simply be directed at constructing lines of defense against the possible. Our experience with nuclear practices would seem to bear this out. But the response has been a renewed effort to further define the possible (more refined techniques of risk analysis, probability assessment, etc.). From this point of view, the burial in the desert amounts to the extension of thinking about the possible to its absolute temporal limits. Yet what is excluded from this picture is everything that threat *can* accomplish that does not resemble what its possibilities are thought to include.

What kind of relationship might we choose to have with the traumatic reality of nuclear waste, and nuclear and ecological threats generally? The approach taken to the waste aspect of this problem has been equally frightening and fascinating. In one sense, it has been grasped as a problem of projecting meaning; a problem of historical transmission: building a sign that would retain its distinctive features qua sign for 10,000 years. But even with diverse inputs into the development of the sign, it has remained essentially a *technical* question. A question of design. A question of building a better sign. Better meaning. And it strikes me that in exactly the same way as the physical materials are seen to be technical questions—the materials that must be developed to withstand a certain average wind pressure and temperature, a certain average particulate load, etc.—so it is that the sign has been seen as a question of assembling *better* semiotic units. All of this amounts to building a sign that can *shout* louder in order that it can *mean* longer. There is an interesting equivocation going on in relation to the idea of distance; the clarity of the sign in its spatial aspects is taken to be a guarantor of the sign's clarity in its temporal aspects. And then there is the double movement of the burial in which the whole question of the monument is sealed. On the one hand, the waste is to be interred, buried, and thus concealed, made *safe*. And on the other hand, the very danger is to be made manifest again through signification. The presumption is an intimate bond (signifier = signified), an eternal motivation, between the signified and the signifier, the danger and the sign, burial and signification. The wager is that the *proper* signifier will retain this integrity and not engage in a kind of nuclear-mimetic deterioration with its signified (signifier < signified). The proper sign is presumed to possess a perpetual and veridical iconicity together with a perfect exclamatory indexicality. It is a perfectly naïve view of meaning in which, as Foucault puts it, things "murmur meanings our language merely has to extract."[106] But here, signification is and must be the only game in town (even, and particularly after, the town is long gone).

With the monument as an anchor dropped hopefully in the ontic, tenuously attached to the present with the anxiety of a profound uncertainty, the problem has found the solution it deserved. Somehow though, perhaps through overcommitment to the episteme of "risk," together with an unstated wish for a redemption, those charged with responsibility for this problem have opted to dispense with memory and focus instead on the unprecedented expenditure of the burial and the monument.

Through the use of a monument, the task of memory is symbolically deposited within a sign that is then given the task of organizing this meaning through time. The presence of threat is shifted away from the concreteness of human practice, to monumentality. Through repression, denial, and disavowal the problem is made to reach closure. Yet this closure is entirely premature—if it is even closure that is called for—since no one has been able to come to terms with what has taken place.

The gravestone has become the only real issue. Yet the question of the burial has never been posed. Why bury? It is an odd fascination with a certain function of death, with making these materials die. Threat, though, cannot be provoked into death in this fashion. And the whole problem with these nuclear materials is that they—and the threat they pose—refuse to die to begin with. Yet somehow we are driven to perform the last rites well in advance—it is one thing to mark our wish for their hasty demise, but building the coffin and cutting the stone borders on a nearly unimaginable disavowal.

Threat, though, must force us to confront the question of the monument, of death, of semiosis, and of a cultural otherness that confronts us when we are drawn to think of the future in this manner. But, from the point of view of the marker, such considerations might also reverse the manner in which the monument is to be thought. That is, rather than the double movement of the burial, it might allow us to talk precisely about signs that hurt. Reliance upon a foggy likeness of danger amounts to a hope for an eternal firstness—ever ready to encounter the secondness of a body—for an eternal expression of the possible, *regardless of aught else*. Instead, the project might be drawn to consider that the materials should be made not "safe" but as dangerous as they really are.[107] Not signification, but knowledge and memory and practice in relation to dangerous material.

Dread

Men make their own history, but they do not make it just as they please; they do not make it under circumstances chosen by themselves, but under circumstances directly encountered, given, and transmitted from the past. The tradition of all the dead generations weighs like a nightmare on the brain of the living.

Karl Marx, "The Eighteenth Brumaire
of Louis Bonaparte"

I wonder what one's body would feel like as it stands in the desert in the presence of such a monument, knowing it to be a place like no other. How can one respond to the singularity of the place, to the negative intention of the monument (except by being there)? The figure in the drawing (Figure 17) is doing what most would do, I

Figure 17. Landscape of Thorns. Concept by Michael Brill and art by Safdar Abidi.

think. The figure is in a pose that recalls a contemplation. What else? This is an entirely conventional expectation of what monuments do. Contemplation, and perhaps an awe. But contemplative is the right word only if we keep in mind its nonmystical connotation. It is a bringing together of a sense of viewing or witnessing strictly, with that of a *templum*—"a site marked out for observation of the auguries." Together it is a site and a practice. A site of memory and a site of awe. But in this case the issue *is* the memory. Which is to say, a forgetfulness. I cannot remember, and even if I can, it is a memory that is singular and particular, not social, deliberate, not collective. Contemplation must run aground in the site of memory where the memory itself is absent. This is not a *lieu de mémoire*. In this sense, the contemplative posture of the human witnesses is precisely wrong. The contemplative posture finds its home in a commemoration. What must be recalled here, what must be rendered sensible by the monument is an utterly singular past. And regardless of whatever else one might like to attribute to it, it is a past that must be understood as dangerous (literally) to the present. It is a past that persists into the present, that impinges on us in a way that makes concrete that which was symbolic or virtuous: the dependence of the present upon the respectful understanding of the past. Not a past that is "significant" to the present, nor a past that is "worthy" of recollection insofar as it is understood to be the laboratory of the present. Rather, a past that *actually* persists in the form of a threat to the present.

What is concealed in the image, like the waste that lies concealed beneath, is perhaps closer to dread. Kierkegaard's dread—as an anxiety with respect to the future—is present here in the realization that the past persists only in the possibility of its repetition—i.e., the future.[108] It is because the past has not been allowed to *be past* that dread or anxiety is produced (for if the past were really past, one could not feel dread, but only repentant, says Kierkegaard). The very possibility that the past invoked by the monument can persist and repeat into the future is what realizes dread. Dread of and for the future as the past's repetition.

What would it be like to stand as the little human drawings in front of something intended to last longer than anything built in Alexander's time or before? The canonical wonders, the Seven Wonders, are all but gone, salvaged, stolen, and sold by raiders, builders, and collectors, and traders, and archaeologists. Fragments, it is said, are in foundations, in walls, in pottery and poetry, undiscovered and underground, and in the British Museum. The pyramid of Khufu at Giza, built a mere 4500 years ago, is still there (except for all the nice white siding, also salvaged over the years). Nebuchadnezzar's hanging gardens are mostly the stuff of fable. Earthquakes brought down the Colossus at Rhodes and the lighthouse at Pharos. Bits of the lighthouse were eventually used to construct a garrison at Alexandria. And some 900 years after it fell, the Colossus was cut up and sold to a Syrian scrap dealer (the oracle had cautioned the Rhodians against reconstructing it.) The statue of Zeus at Olympia, damaged by Caligula's workers, then relocated to Constantinople, was eventually destroyed in a fire in AD 462. A few pieces of the Temple of Artemis still exist, having endured arson, sacking by Goths, and the eventual dismantling by St. John Chrysostom.

Some say that all the Seven Wonders existed in time for a period of only forty years. One could have traveled the Hellenistic universe of the great Alexander and have seen all the wonders of the world. All of them. The British architect Paul Shepheard has this phantasy. He figures it would have taken about six months to do it. I'm not so sure.

The monument must always work this way. Either forgetting to remember or remembering to forget. It can only allow the past to be present in the form of a threat to life. The penitence cannot fully consign the threat to a past, to any past. This is the problem. Fear. And trembling. Writes Derrida: "We tremble in that strange repetition that ties an irrefutable past (a shock has been felt, a traumatism has already affected us) to a future that cannot be anticipated....I tremble at what exceeds my seeing and my knowing...a secret always makes you tremble."[109] The auguries are not very favorable. Both, that is, in the present sense—that *of* a future—and in the sense that the monument (a monument) is to carry the burden of memory of that which cannot be recalled. The recollection of the unrecallable in the face of the incomprehensible.

A material trace, the visibility of the monument, supported not by a collective and encompassing memory, but by an archival assemblage: you are here.

Appendix: A WIPP Chronology

1955   The U.S. Atomic Energy Commission asks the National Academy of Sciences to study permanent disposal of radioactive wastes. The following year, the academy recommends disposal in salt deposits.

1970   Based on testing done in the 1960s by the Oak Ridge National Laboratory (ORNL), a Lyons, Kansas, salt mine is selected as a potential site for a radioactive waste repository.

1972   The Lyons site is judged unacceptable because of the area's geology, hydrology, and previously undiscovered drill holes that could lead to extensive dissolution of salt.

1974   A site thirty miles east of Carlsbad, New Mexico, is chosen for exploratory work. Several boreholes are drilled by ORNL.

1975   A borehole drilled to three thousand feet provides enough information to eliminate the original site. Severe structural deformations in the primary salt bed and pressurized brine hundreds of feet deeper were factors. The U.S. Geologic Survey recommends moving to an area about six miles southwest of the original borehole.

1978   The Environmental Evaluation Group (EEG) is established to provide the State of New Mexico with oversight powers.

1979   Congress authorizes the Waste Isolation Pilot Plant (WIPP) for the research and development of safe methods of disposal of radioactive wastes generated by defense facilities. Against President Carter's recommendations, Congress authorizes WIPP for defense-related transuranic waste (TRU) only, and exempts it from oversight and licensing of the Nuclear Regulatory Commission on security grounds.

1980   The Department of Energy (DOE) issues its Final Environmental Impact Statement on the WIPP declaring the site appropriate for construction.

1981   New Mexico Attorney General sues the DOE and the Interior Department. The suit is resolved by an agreement requiring more studies and guaranteeing the state more information. The first exploratory shaft is drilled.

1982   Underground excavation begins. The DOE and the State of New Mexico sign an agreement committing the DOE to seek money from Congress for upgrading WIPP transportation routes and clarifying DOE liability for accidents at the WIPP or en route.

1983   The DOE decides to proceed with full facility construction of the WIPP.

1984   The DOE and the State of New Mexico agree that the WIPP must comply with all state, federal, and local laws and regulations, including those of the Environmental Protection Agency (EPA).

1985   The EEG notifies the DOE that the original container proposed for WIPP shipments is unacceptable. The EPA establishes radioactive waste disposal regulations applicable to the WIPP.

1986   The EPA says the WIPP must comply with the Resource Conservation and Recovery Act (RCRA) of 1976 for disposing of mixed hazardous and radioactive waste.

1987   The DOE selects a new design for the shipping containers. A federal court invalidates part of the EPA's radioactive waste disposal rules, leaving no permanent repository regulations applicable to WIPP. A modified agreement between the DOE and New Mexico commits WIPP to original rules until revised regulations may be put in place.

1988   The DOE announces WIPP will not open as scheduled.

1989   The DOE applies to the Interior Department for the withdrawal from public use of 10,240 acres of federal land surrounding the site on which and under which WIPP is constructed. The DOE petitions the EPA for a waiver from RCRA land disposal restrictions. The DOE issues its five-year test plan for WIPP. The Nuclear Regulatory Commission approves new transportation containers. The DOE proposes July 1, 1990, as the earliest possible WIPP opening date.

1990   The DOE issues its Supplemental Environmental Impact Statement on WIPP. The EPA grants a No Migration Determination, meaning that no hazardous materials would escape from the repository to the environment during the planned test phase. The DOE issues its revised plan for a five-year test phase at WIPP, including gas-generation experiments. The EPA authorizes the New Mexico Environmental Improvement Division (now the Environment Department) to regulate disposal of mixed radioactive and hazardous waste, including the WIPP waste.

1991   A complicated year. The Interior Department administratively transfers WIPP land to the DOE; the House Interior and Insular Affairs Committee passes a resolution nullifying the administrative transfer; New Mexico files suit against the DOE claiming that the land transfer must be made by Congress. The suit is joined by the State of Texas and several members of Congress; the U.S. Senate passes, by unanimous consent, the interdepartmental transfer of WIPP land as proposed by New Mexico's U.S. Senators, Republican Pete Domenici and Democrat Jeff Bingaman.

1992   A permanent injunction on the WIPP site is granted (later overturned), and the RCRA permit is invalidated by U.S. District Court Judge John Garrett Penn in Washington, DC; Congress passes and President Bush signs the WIPP "Land Withdrawal Act."

1993   In response to opposition to any on-site testing, the DOE claims that tests with radioactive waste at the WIPP site are not necessary to show compliance. This, as Jeff Wheelwright points out in "For Our Nuclear Wastes, There's Gridlock on the Road to the Dump" (*Smithsonian* 26, no. 2 [1995]), was a symbolic and tactical loss for WIPP; had on-site testing been pursued, the DOE would have arguably surmounted the additional hurdle of actually transporting waste through the state and placing it in the site. The Concerned Citizens for Nuclear Safety (a stakeholder organization in the Southwest) puts a different spin on this decision. They suggest that it was the reticence on the part of the DOE to submit the site to actual testing and the scrutiny of the EPA that motivated the decision to conduct computer modeling tests only. (See Concerned Citizens for Nuclear Safety, "WIPP—Where Do We Stand?" [1996, http://www.nuclearactive.org/wipp/index.html].) NMED issues a draft RCRA permit, indicating a final permit may be issued in 1994 to allow WIPP to open following a period of public comment and hearings; the DOE declares readiness to begin receiving waste for a test phase; the DOE announces it will do radioactive testing at a national laboratory rather than at the WIPP site. Accelerated regulatory compliance becomes the main focus at the site; the DOE appoints George E. Dials as manager of a newly created Carlsbad Area Office, linking the WIPP directly with DOE Headquarters in Washington, DC.

1994   Energy Secretary Hazel O'Leary visits the site, reaffirming her commitment to open the WIPP. She returns on October 6 to present the DOE's first Voluntary Protection Plan "Star" Award to the WIPP.

1995   The DOE submits a Draft Compliance Certification Application (DCCA) to the U.S. Environmental Protection Agency. The DOE submits a revised Resource Conservation and Recovery Act (RCRA) permit application to the New Mexico Environment Department. The DOE submits to the EPA a draft petition for a No-Migration Variance Determination. The WIPP Disposal Decision Plan is revised, accelerating the opening by two months. The Energy Secretary's disposal decision is moved up to October 1997, and the first receipt of waste is scheduled for April 1998.

1996   The DOE submits the *Disposal Phase Draft Supplemental Environmental Impact Statement (SEIS–II)*, which analyzes the potential impacts of disposing TRU at the WIPP. In this document, six scenarios are assessed for the WIPP: the proposed action is to continue as planned, and the alternatives are to use WIPP for the disposal of defense and nondefense waste; the disposal of treated (to reduce volume) TRU as well as other wastes deemed appropriate by the DOE (nondefense related); the disposal of TRU after treating it by a "shred and grout" process (i.e., grinding and mixing waste with material that hardens and immobilizes the contaminants); do nothing, and instead close the WIPP and deal with wastes by temporary storage in monitored sites; do nothing, and close the WIPP, leaving existing waste where it is and dealing only with newly generated TRU. Interestingly, the EPA grants an exemption to the "no migration" clause that is specified in the RCRA. The EPA states that the requirement "is redundant to the more stringent radioactive waste disposal criteria, and that the exemption will not jeopardize the environment nor the public health and safety"(WIPP press release, "President Signs Opening of theWIPP, Save Taxpayers Money Bill to Speed" [September 23, 1996]; U.S. Department of Energy [1996], http://www.wipp.carlsbad.nm.us/pr/prindex .htm). Two other significant features of 1996 were the issuing of the compliance criteria to which WIPP would conform (otherwise known as 40 CFR 194, Criteria for the Certification and Re-certification of the Waste Isolation Pilot Plant's Compliance with the 40 CFR Part 191, Environmental Radiation Protection Standards for the Management and Disposal of Spent Nuclear Fuel, High-Level and Transuranic Radioactive Waste), and the response submitted by DOE/WIPP to these criteria: the massive Compliance Certification Application. This document, representing the ultimate textual potlatch, is really too vast to read; it spans eighty-four thousand pages of text, appendixes, figures, tables, and references. The EPA requires a year to evaluate it. From the point of view of the interested party, the document is doubly difficult to read. First, paper copies do not exist for circulation, and second, CD-ROM versions exist in only eleven locations in the Southwest United States.

1997   In March, the DOE revises WIPP's opening date to May 1998. The EPA informs the DOE that it would not complete the required rule making on the certification decision until April 1998 at the earliest. The *Albuquerque Journal* reports on April 1, 1997, that the delay was actually due to time needed to verify computer models of the WIPP that show that the site will be able to prevent the waste from reaching the accessible environment for ten thousand years. In a ten-year plan for the site, two scenarios are given, one in which the WIPP would receive waste from only a small number of sites, and scale operation to a period of one hundred years, and the other that the site would operate at full capacity, ceasing operations after thirty-five years (U.S. Department of Energy, "Ten-Year Plan Looks toward the Future," *TRU Progress* [winter 1997], http://www.wipp.carlsbad.nm.us/pr/truprog/TRUmenu.htm).

1999   On March 26, Energy Secretary Bill Richardson announces that the first shipment of defense-generated TRU arrived safely at the WIPP.

Notes

Preface

1. See James Lovelock, *Gaia: A New Look at Life on Earth* (New York: Oxford University Press, 1979); Barry Commoner, *The Closing Circle* (New York: Bantam Books, 1972); Anthony Wilden, "Epistemology and Ecology: The Difference That Makes the Difference," in *System and Structure: Essays in Communication and Exchange* (New York: Tavistock, 1980); and Verena Andermatt Conley, *Ecopolitics: The Environment in Poststructuralist Thought* (New York: Routledge, 1997).

2. See, for example, Theodore Rozak, *Where the Wasteland Ends* (Garden City: Doubleday, 1972), 400.

3. Rachel Carson, *Silent Spring* (New York: Houghton Mifflin, 1962).

4. Félix Guattari, *Chaosmosis: An Ethico-aesthetic Paradigm*, trans. Paul Bains and Julian Pefanis (Bloomington: Indiana University Press, 1995), 91. In "The Three Ecologies" (*New Formations* 8 [1989]: 131–47), Guattari called for a slightly different sense of ecology: "The generalized ecology I am arguing for here has in my view barely begun to be prefigured by environmental ecology in its contemporary form. . . . Existing ecological movements certainly have many virtues; but the wider ecological question seems to me too important to be abandoned to the archaizing, folkloristic tendencies which choose determinedly to reject large-scale political involvement . . . for the ecology I propose here questions the whole of subjectivity and capitalist power formations" (140).

5. Sigmund Freud, "The Uncanny" (1919), in *Collected Papers*, Volume 4, trans. Joan Riviere (New York: Basic Books, 1959), 399.

6. Peter C. van Wyck, *Primitives in the Wilderness: Deep Ecology and the Missing Human Subject* (Albany: State University of New York Press, 1997).

7. "A Klee painting named 'Angelus Novus' shows an angel looking as though he is about to move away from something he is fixedly contemplating. His eyes are staring, his mouth is open, his wings are spread. This is how one pictures the angel of history. His face is turned toward the past. Where we perceive a chain of events, he sees one single catastrophe which keeps piling wreckage and hurls it in front of his feet. The angel would like to stay, awaken the dead, and make whole what has been smashed. But a storm is blowing in from Paradise; it has got caught in his wings with such a violence that the angel can no longer close them. The storm irresistibly propels him into the future to which his back is turned, while the pile of debris before him grows skyward. This storm is what we call progress." Walter Benjamin, "Theses on the Philosophy of History," in *Illuminations*, ed. Hannah Arendt, trans. Harry Zohn (New York: Schocken Books, 1968), section 9.

8. Alan Burdick, "The Last Cold-War Monument: Designing the Keep Out Sign for a Nuclear-Waste Site," *Harper's*, 289, no. 8 (1992): 62–67.

Introduction

1. Press release, U.S. Department of Energy, March 26, 1999.

2. I will return to this theme later. See also Paul Virilio, "The Museum of Accidents," *Public* 2 (1989): 81–5; Paul Virilio, "The Primal Accident," in *The Politics of Everyday Fear*, ed. and trans. Brian Massumi (Minneapolis: University of Minnesota Press, 1993); Paul Virilio, *Open Sky*, trans. Julie Rose (New York: Verso, 1997), 69–86.

Waste

1. Jody Baker, "Modeling Industrial Thresholds: Waste at the Confluence of Social and Ecological Turbulence," *Cultronix* 1, no. 1 (1994). http://eserver.org/cultronix/baker. The two series can be pictured thus: Waste / space / containment, and Decay / time / disposal.

2. The situation is more complex than this. The actual isotope of uranium that is necessary to sustain a reaction (U-235) in a typical American reactor is in fact a very small percentage of "natural" uranium (U-238), somewhat less than 1 percent. During the 1970s it was thought that global supplies of in situ uranium were quite low. However, a byproduct of the nonfissile U-238 is plutonium—itself a suitable reactor fuel. Thus the non-fuel in natural uranium produces a fuel source: this is the basis of the concept of a breeder reactor. And its promise is near unlimited fuel from a limited resource. The downside, of course, is that plutonium is also the ideal source for the manufacture of weapons.

3. Writes Linda Rothstein: "The tank farms at the Hanford Site hold 61 million gallons of liquids and sludges. The contents include radioactive waste and spent fuel from nine weapons production reactors mixed with assorted hazardous chemicals, including nitrates and nitrites, chromium, mercury, and cyanide. By the early 1990s, 24 of the tanks were considered in some danger of exploding, according to William Alumkal, who is the Executive Vice President of Westinghouse Hanford's tank waste remediation division." Linda Rothstein, "Nothing Clean about 'Cleanup,'" *Bulletin of the Atomic Scientists* 51, no. 3 (1995), http://www.bullatomsci.org/issues/1995/mj95/mj95.rothstein.html. See also the Hanford Health Information Archives, http://www.hhia.org/; Witness to the Future, http://witnesstothefuture.com/text/main.html; Hanford Holiday, http://www.thestranger.com/1999-02-04/index.html.

4. The following was reported in *Covert Action:* On January 14, 1991, just days before the Gulf War, Part I, the Pentagon leaked to *Newsweek* General Norman Schwarzkopf's request for permission to use nuclear weapons in the Gulf. It is also reported that there were approximately 300 U.S. hydrogen bombs in the Gulf aboard U.S. ships. Michio Kaku, "Nuclear Threats and the New World Order," 1992, http://www.ratical.org/radiation/inetSeries/NthrtsNnwo.html.

5. The newer generation of weapons, "'brilliant' weapons find their own targets, unlike mere smart bombs, which have to be directed. One example is Sadarm (Seek And Destroy Armor). It is fired like a normal artillery shell into the target area, where it ejects two submunitions that descend by parachute. As they fall, Sadarm scans the ground with radar and infrared sensors.... Sadarm fires a slug of molten metal at the selected target. The slug takes on an aerodynamic shape as it travels through the air, ideal for piercing armor. Though less powerful than a DU shell, it can break through the top armor of any tank.... Sadarm can be lobbed at an enemy 20km away.... Sadarm and other brilliant weapons use tantalum, an exotic heavy metal for which little data is available. But it appears to be highly toxic, especially when vaporized." See David Hambling, "Why Deadly Depleted Uranium Is the Tank Buster's Weapon of Choice," *Guardian Weekly*, May 18, 2000.

6. See Mary Douglas, *Purity and Danger: An Analysis of the Concepts of Pollution and Taboo* (London: Arc, 1966).

7. Since the 1986 Chernobyl accident, cesium fallout is precisely the problem in the regions surrounding the plant. In areas of Belarus as far as 350 km from the nuclear plant, cesium contamination has resulted in staggering incidences of cancers among children (particularly thyroid). The pathways of contamination vary from simply drinking water to fueling baker's ovens with irradiated wood.

8. See Georges Bataille, "Concerning the Accounts Given by the Residents of Hiroshima," in *Trauma: Explorations in Memory*, ed. Cathy Caruth (Baltimore: Johns Hopkins University Press, 1995).

9. See Ian Hacking, *The Social Construction of What?* (Cambridge: Harvard University Press, 1999).

10. It is odd, in a way, that these are the *meaningful* reference points. The veiled history of nuclear weapons testing in the United States is a terrifying piece of history. Two extraordinary books that document this history of secrecy around U.S. nuclear and conventional weapons testing are Richard Misrach and Myriam Weisang Misrach, *Bravo 20: The Bombing of the American West* (Baltimore: Johns Hopkins University Press, 1990); and Carole Gallagher, *American Ground Zero: The Secret Nuclear War* (New York: Random House, 1993). The former documents the history of the U.S. Navy's control of over 70 percent of Nevada's airspace and its bomb testing. The latter is a collection of photographs and personal narratives documenting the nuclear testing in Nevada, Utah, and Arizona.

11. Rosalie Bertell, *No Immediate Danger: Prognosis for a Radioactive Earth* (Toronto: Women's Educational Press, 1985).

12. James Megaw, *How Safe? Three Mile Island, Chernobyl, and Beyond* (Toronto: Stoddart, 1987), 156–67.

13. Bertell, *No Immediate Danger*, 206.

14. Incidentally, in my copy of the *International Nuclear Fuel Cycle Fact Book* (Washington, DC: U.S. Department of Energy, 1992)—essentially a global telephone book for nuclear installations and waste management facilities—the Mayak facility is, out of hundreds of entries covering twenty-three countries, the only facility that has neither phone nor fax numbers listed.

15. This area has been the subject of a recent documentary: *Chelyabinsk: The Most Contaminated Spot on the Planet*, 1995, by the American-Polish director Slawomir Grunberg.

16. See Ghost Town—Chernobyl Pictures—Elena's Motorcycle Ride Through Chernobyl, www.kiddofspeed .com. See also Valentin Tikhonov, ed., *Russia's Nuclear and Missile Complex: The Human Factor in Proliferation* (Washington, DC: Carnegie Endowment for International Peace, 2001); and Jon Brook Wolfsthal, Cristina-Astrid Chuen, and Emily Ewell Daughtry, eds., *Nuclear Status Report: Nuclear Weapons, Fissile Material, and Export Controls in the Former Soviet Union* (Washington, DC: Carnegie Endowment for International Peace, 2001). Both are available at the Carnegie Web site.

17. Quoted in Spencer Weart, *Nuclear Fear: A History of Images* (Cambridge: Harvard University Press, 1988), 371.

18. For a detailed account of the health-related fallout of Three Mile Island as well as a strongly worded warning that should something similar happen, absolutely nothing will have been learned from these events, see Harvey Wasserman and Norman Solomon, *Killing Our Own: The Disaster of America's Experience with Atomic Radiation* (New York: Dell, 1982), http:// www.ratical.com/radiation/KillingOurOwn/.

19. These events are now part of the public record. I have relied on a number of sources for this gloss. Particularly good are Mike Gray and Ira Rosen, *The Warning: Accident at Three Mile Island* (Chicago: Contemporary Books, 1983); Megaw, *How Safe?*; Charles Perrow, *Normal Accidents: Living with High-Risk Technologies* (New York: Basic Books, 1984); Richard Wolfson, *Nuclear Choices: A Citizen's Guide to Nuclear Technology* (New York: McGraw-Hill, 1991).

20. This failure mode is a classic example of what John Gall terms "wishful feedback." See John Gall, *Systemantics: The Underground Text of Systems Lore: How Systems Really Work and How They Fail* (Ann Arbor: General Systemantics Press, 1986), 140.

21. This manner of accident is hardly novel. See Perrow's *Normal Accidents* for other examples of "coupling"-type accidents.

22. Virilio speaks in these terms in a number of places. See Paul Virilio, *Territorial Insecurity* (Minneapolis: University of Minnesota Press, forthcoming); Paul Virilio, "The Accident Museum," in *A Landscape of Events*, trans. Julie Rose (Cambridge: MIT Press, 2000), 54–60; Paul Virilio, "The Primal Accident," in *The Politics of Everyday Fear*, ed. and trans. Brian Massumi (Minneapolis: University of Minnesota Press, 1993); Paul Virilio, "Surfing the Accident: An Interview with Andreas Ruby," in *The Art of the Accident*, ed. Andreas Broeckmann, Joke Brouwer, Bart Lootsma, Arjen Mulder, and Lars Spuybroek (Rotterdam: NAi Publishers/V2_Organisatie, 1998).

23. Octavio Paz, "Order and Accident," in *Conjunctions and Disjunctions*, trans. Helen R. Lane (New York: Viking Press, 1974), 111–12.

24. Ibid., 112.

25. See Perrow, *Normal Accidents*, 15–31.

26. Kai T. Erikson, *A New Species of Trouble: The Human Experience of Modern Disasters* (New York: Norton, 1994), 194. It is an important dimension for Erikson's work on modern disaster that there is no clear ending in the disastrous events. "The feeling of uncertainty—the lack of a sense of ending—can begin the very moment that the event ought, in logic, to be over" (148). And I also think that the indeterminacy of beginnings is of equal importance.

27. To be fair, I suppose, to the category of *catastrophes of the inevitable* we should probably add another species of inevitability—*accidents of the intentional*. For example, from a 1992 report on nuclear accidents: "The Soviets revealed that they have been dumping radioactivity into the Kara Sea, which connects to the Arctic Ocean, for three decades. Besides four nuclear-powered submarines lost at sea, the Soviets said they dumped four decommissioned naval nuclear reactors in 1965 and 1966, three reactors from the icebreaker Lenin in 1967, a barge carrying a submarine reactor sunk in 1972, and a nuclear-powered submarine jettisoned its reactor core in 1982. Charles Hollister of the Woods Hole Oceanographic Institution calculates that the Soviets dumped about 600 million curies of radioactivity into the ocean, or roughly seven times as much radioactivity as was in the Chernobyl reactor that melted down April 26, 1986." Peter Montague, "The Year in Review: Nuclear," *Rachel's Hazardous Waste News* 317 (1992), http://www.rachel.org/ bulletin/bulletin.cfm?Issue_ID=812. However, this is not to suggest that the (then) Soviet Union has (or had) a proprietary relationship with carelessness. In the same year as this report was released, the U.S. Army admitted having inadvertently shipped a kilogram of plutonium via Federal Express. *Facts on File World News Digest*, Dec. 22, 1994, cited in Peter Montague, "The Fourth Horseman: Nuclear," *Rachel's Environment and Health*

Weekly 473 (1995), http://www.rachel.org/bulletin/
bulletin.cfm?Issue_ID=650.

28. An "action-goading fear." See Weart, *Nuclear Fear.*

29. Ibid., 305–6. Weart's position is that the anxiety
provoked in relation to nuclear practice is an imaginary
response—a nuclear fear, as he calls it. In other words
the actual record of, for example, chemical or mining
industries or damming projects provides a much more
"reasonable" grounds for anxiety concerning big science
and technology. Yet one could only reach this conclusion
by considering such threats objectively, as elements of
actual risk. He fails to see that the reason for the shared
imaginary around nuclear threat—out of proportion to
its objective body count as it may be—is the result of the
nuclear itself. However, my interest here should not be
seen as a privileging of the nuclear in relation to ecological threat. Without a doubt, the contingency and
potential consequences of events such as Bhopal, India;
Buffalo Creek, West Virginia; and Mississauga, Ontario,
contribute to a social climate in which these threats and
incidents (or representations of them) are internalized.

30. See Jean Laplanche and J.-B. Pontalis, *The Language
of Psychoanalysis,* trans. Donald Nicholson-Smith (New
York: Norton, 1973), 97–103.

31. The periodic table has a rather fascinating history.
See J. W. van Spronsen, *The Periodic System of Chemical
Elements: A History of the First Hundred Years* (Amster-
dam: Elsevier, 1969); J. S. F. Pode, *The Periodic Table:
Experiment and Theory* (New York: Wiley, 1971); J. R.
Partington, *A History of Chemistry,* vol. 4 (New York:
Macmillan, 1964); W. H. Brock (ed.), *The Norton History
of Chemistry* (New York: Norton, 1993).

32. The "new" elements 118 (ununoctium) and 116
(ununhexium) were made by accelerating a beam of
krypton-86 ($^{86}_{36}$Kr) ions (to an energy of 449 million
electron volts) toward a target of lead-208 ($^{208}_{82}$Pb) for
some 300 hours. A total of three atoms of the new elements were identified—both of which have half-lives
of less than a millisecond. The rate of production of
ununoctium is roughly 1 atom for every 1012 interactions. Under "Uses" www.webelements.com has the
following entry: "Element 118 has no uses as only about
three atoms of this element have ever been made."

33. The following discussion relies mostly on my memory of physics and chemistry classes. There are numerous
basic sources for this sort of information. Two reasonably good ones are Wolfson, *Nuclear Choices;* and League
of Women Voters, *The Nuclear Waste Primer: A Handbook
for Citizens* (New York: Lyons and Burford, 1993).

34. The following (a kind of Cold War rumba or apocalyptic jingle) was a popular education mantra of the late
1950s. The response to threat was discipline, responsibility, and always, and above all, alertness: "Deetle dum
dum, deetle dum dum, / There was a turtle by the name
of Bert, / and Bert the Turtle was very alert. / When
danger threatened him he never got hurt, / he knew just
what to do." Chorus: "Duck, and Cover./ Duck, and
Cover. / He did what we all must learn to do, / you and
you and you and you: Duck, and Cover." Voice-over:
"Now, you and I don't have shells to crawl into like Bert
the Turtle, so we have to cover up in our own way. Paul
and Patty know this, no matter where they go, or what
they do, they always try to remember what to do if the
atom bomb explodes right then. *It's a Bomb! Duck! and
Cover!* Now here's Tony going to his Cub Scout meeting.
Tony knows that the bomb can explode any time of the
year, day or night. *Duck! and Cover! Ataboy, Tony, that
flash means act fast!*" Federal Civil Defense Administration, *Duck and Cover,* film (Archer Production
Incorporated, Washington, DC, 1955).

35. Emergency Measures Organization, *11 Steps for
Survival: Blueprint for Survival No 4* (Ottawa: Queens
Printer, 1961).

36. See, for example, Office of Environmental Man-
agement, "Linking Legacies," 58, 80; and Office of
Environmental Management, "Accelerating Cleanup:
Focus on 2006," discussion draft, DOE/EM-0342
(Washington, DC, June 1997), E-1.

37. A list of several countries, the earliest likely date for
a working disposal system, and type of facility is as follows: Belgium, 2020, underground laboratory in clay at
Mol; Canada, no date, after an independent commission
conducted a four-year study concerning the burial of
irradiated fuel in granite, the project was shelved; China,
no date, irradiated fuel to be reprocessed, Gobi desert
sites under investigation; Finland, 2020, field studies
being conducted; France, 2010, two sites to be selected
for study, final selection by 2006; Germany, 2008,
Gorleben salt dome sole site to be studied; India, 2040,
irradiated fuel to be reprocessed, waste stored for 20
years in yet to be identified granite site; Italy, 2040, irradiated fuel to be reprocessed and stored for 50 to 60
years and then buried in clay or granite; Japan, 2020,
limited site studies, cooperative program with China to
build underground facility; Netherlands, 2040, interim
storage of reprocessed waste for 50 to 100 years before
burial; Russia, no date, program uncertain, although they
are now actively importing nuclear waste; Spain, 2020,
burial in unidentified clay, granite, or salt formation;
Sweden, 2015, granite site to be selected in 2008; United
States, 2010, Yucca Mountain, Nevada, given approval in
2002, and if actually developed, will receive 70,000 tons
of high-level waste; United Kingdom, 2030, 50-year storage approved in 1982, long-term options include seabed
burial. Adapted and updated from "Nuclear Waste: The
Problem That Won't Go Away," *Worldwatch Institute,*

December 1991, 24–25; and U.S. Department of Energy and Office of Civilian Radioactive Management, *International Fuel Cycle Fact Book* (Oak Ridge, TN: U.S. Department of Energy, DOE/RW-0371P, 1992).

38. See Robert Alvarez, "Energy in Decay," *Bulletin of the Atomic Scientists* 56, no. 3 (2000): 24–35.

39. See http://www.skullvalleygoshutes.org/. The Skull Valley Goshutes are already disappearing. Located in Tooele County—which has the distinction of being recognized as the most toxic county in the United States—there are fewer than thirty band members still living on the reservation.

40. Kathleen M. Trauth, Stephen C. Hora, and Robert V. Guzowski, *Expert Judgment on Markers to Deter Inadvertent Human Intrusion into the Waste Isolation Pilot Plant* (Albuquerque, NM: Sandia National Laboratories, 1993), 1–1. For a longer chronology of the WIPP, see appendix A.

41. The methodological process of "expert-judgment analysis" was used extensively throughout the development of the project. It is a method of extracting probability estimates with respect to questions that have significant and unresolvable uncertainties. Essentially this means assembling a group of experts, dividing that group into teams, assigning a set of rules around the assignment of probability estimates, defining a clear statement of the issue to be judged, and then proceeding to reduce the issue to dimensions that can be conceptually assigned to a finite probability scale. See E. J. Bonano, S. C. Hora, R. L. Keeney, and D. Winterfeldt, *Elicitation and the Use of Expert Judgment in Performance Assessment for High Level Radioactive Waste Repositories* (Albuquerque, NM: Sandia National Laboratories, 1990).

42. Throughout the testimony given at the EPA public hearings in Santa Fe, Albuquerque, and Carlsbad for certification of the WIPP, many concerns were raised with respect to the mining techniques that were assumed in the DOE compliance data. Air drilling and drilling with light fluids were not considered. The complete testimony is available at http://www.epa.gov/radiation/wipp/docs/.

43. Jean Baudrillard, *America*, trans. Chris Turner (London: Verso, 1988), 123–24.

44. Ibid.

45. See Gary Genosko, *Baudrillard and Signs: Signification Ablaze* (New York: Routledge, 1994), 117–29.

46. Baudrillard, *America*, 123.

47. Gilles Deleuze and Félix Guattari, *A Thousand Plateaus*, trans. Brian Massumi (Minneapolis: University of Minnesota Press, 1987), 228.

48. Ibid.

49. Ibid., 287.

50. Jean Baudrillard, *Cool Memories II*, trans. Chris Turner (Durham, NC: Duke University Press, 1996), 14–15. This is not, however, to align Baudrillard with Deleuze and Guattari with respect to the secret. Elsewhere—particularly Jean Baudrillard, *Seduction*, trans. Brian Singer (Montréal: New World Perspectives, 1990)—he makes it clear that the secret is a kind of pact in which the content of the secret is secondary or even incidental to the keeping of it; that is, to its remaining unspoken (see 79–81).

51. One of the stealth's most embarrassing public disclosures was in the form of decidedly visible images of wreckage broadcast on Serbian television in 1999.

52. This is a somewhat reckless paraphrase from Derrida's thoughts on secrecy in "Tout Autre Est Tout Autre," in Jacques Derrida, *The Gift of Death*, trans. David Wills (Chicago: University of Chicago Press, 1995).

53. It is unclear to me whether the null hypothesis was really an idea or an alibi. After all, it would cost a fraction of the proposed amount to bury the waste in an unmarked site—providing a compelling incentive for the project to be spun in this direction. But on the other hand, doing so would make invisible the very gesture of interment.

Dangerous Signs

1. The stone bore an inscription in two languages and three scripts (Greek, Demotic, and hieroglyphs), and its discovery led to the first decipherment of ancient Egyptian inscription.

2. Carl Sagan, F. D. Drake, Ann Druyan, Timothy Ferris, Jon Lomberg, and Linda Salzman Sagan, eds., *Murmurs of the Earth: The Voyager Interstellar Record* (New York: Random House, 1978), 9.

3. Ibid.

4. As for my little experiment, apart from one family member, none of my original correspondents solved the problem.

5. Cf. Jean François Lyotard, "Can Thought Go on without a Body?" in *Materialities of Communication*, ed. Timothy Lenoir and Hans Ulrich Gumbrecht, trans. Bruce Boone and Lee Hildreth (Stanford: Stanford University Press, 1994).

6. See Paul S. Hudson, "The 'Archaeological Duty' of Thornwell Jacobs: The Oglethorpe Atlanta Crypt of Civilization," *Georgia Historical Society* 75, no. 1 (1991): 121–38.

7. See Thornwell Jacobs, "Today-Tomorrow: Archaeology in A.D. 8113 . . . An Opportunity for the

Generation . . . Preserving Records for Posterity . . . Co-operation Needed," *Scientific American* 155 (November 1936): 260–1; and William Jarvis, "Do Not Open until 8113 AD: The Oglethorpe Crypt and Other Time Capsules," *World's Fair* 5, no. 1 (1985): 1–4.

8. One simply must wonder if this previous experience came up at all while Westinghouse negotiated the on-site contract for operation of the WIPP.

9. See David Hillel Gelerntner, *1939: The Lost World of the Fair* (New York: Free Press, 1995); and John E. Findling (ed.), *Historical Dictionary of World's Fairs and Expositions, 1951–1988* (New York: Greenwood Press, 1990).

10. See Stanley Edgar Hyman and St. Clair McKelway, "Onward and Upward with Business and Science: The Time Capsule," *New Yorker*, December 5 (1953): 194–219.

11. Westinghouse Electric Corporation, *The Book of Record of the Time Capsule of Cupaloy Deemed Capable of Resisting the Effects of Time for Five Thousand Years; Preserving an Account of Universal Achievements, Embedded in the Grounds of the New York World's Fair, 1939* (New York: Westinghouse Electric Corporation, 1938). See also G. Edward Pendray, "The Story of the Time Capsule," *Annual Report of the Board of Regents of the Smithsonian Institution—Showing the Operations, Expenditures, and Condition of the Institution for the Year ending June 30, 1939*, ed. Smithsonian Institution Board of Regents (Washington, DC: Government Printing Office, 1939).

12. In 1964, Westinghouse sank another capsule, also at a New York World's Fair and also in Flushing Meadows. This time it was to update the cultural information for the couple of decades that had elapsed since the 1938 capsule. Here we have *modern* objects: a fifty-star U.S. flag, a Bible, credit cards, and an electronic watch. An automatic camera, and contact lenses. Freeze-dried food, plastic wrap, and a bikini bathing suit. A transistor radio and valium. But also a pocket radiation monitor, a sample of carbon-14, antibiotics, birth control pills, and various bits from the early space program. See William Jarvis, "Time Capsules," in *Encyclopedia of Library and Information Sciences*, ed. Allen Kent, vol. 43, suppl. 8 (New York: Marcel Dekker, 1988).

13. Rarely were the names of detonations so loaded, as it were, as Trinity. Indeed, these names themselves would make an interesting study. In my list of detonations between 1945 and 1994 (which, although it runs some forty-three single-spaced pages, is no doubt rather conservative) there is a decided preference (in U.S. detonations) toward naming after Westernized Indian tribal names, the names of foreign cities, and scientists. See Oklahoma Geological Survey Observatory, "Catalog of 1900+ Known Nuclear Explosions," 1994, http://www.unl.edu:2020/alpha/Catalog_of_Known_Nuclear_Explosions.html.

14. TRU is a separate category from high-level waste (HLW) and low-level waste (LLW). HLW, exceedingly dangerous material, is principally spent radioactive fuel rods from commercial and military reactors. TRU is any waste that is heavier than uranium. And LLW is any waste that does not fall into either of the preceding categories. Specifically, TRU is made up of all radioactive isotopes that have an atomic number greater than that of uranium and a decay rate of greater than 100 nanocuries per gram.

15. "The single best measure of a waste's hazard is its (radio)activity, not its volume or half-life. Thus low-level waste has a high volume but a low activity and for the most part a short half-life. High-level waste packs high activity and long life into a relatively small volume. And though TRU is mostly low-volume, low-activity, its extremely long life magnifies the threat in another dimension." Jeff Wheelwright, "For Our Nuclear Wastes, There's Gridlock on the Road to the Dump," *Smithsonian* 26, no. 2 (1995).

16. The total waste to be disposed of from all the generating and storage sites amounts to about 6 million cubic feet (170,000 cubic meters). The U.S. DOE operates the WIPP, aided by the Westinghouse Electric Corporation, the principal contractor. Wastes are to be shipped to the WIPP from ten major Energy Department sites from South Carolina to Washington State, in something on the order of 40,000 shipments over approximately 35 years. The waste to be interred at WIPP will come from the Rocky Flats Environmental Technology Site, Colorado; Idaho National Engineering Laboratory; Los Alamos National Laboratory, New Mexico; Hanford Site, Washington; Lawrence Livermore National Laboratory, California; Nevada Test Site; Oak Ridge National Laboratory, Tennessee; Savannah River Site, South Carolina; Argonne National Laboratory East, Illinois; and the Mound Site, Ohio. En route from these sites, the waste will be transported through Alabama, Arizona, Arkansas, Georgia, Indiana, Louisiana, Mississippi, Missouri, Oregon, Utah, Texas, and Wyoming. The waste is transported to the WIPP in special, 10-foot-high double-walled stainless steel TRUPACT-II shipping containers. The shipments are monitored by DOE's satellite tracking system.

17. To get to the WIPP from Carlsbad, follow Highway 285 south from Carlsbad, and then go east on Route 31. Make a right on Route 128, and proceed east for approximately 12 miles. Follow the signs for the Waste Isolation Pilot Plant.

18. U.S. Department of Energy, "Backgrounder no. 1: What Is the Waste Isolation Pilot Plant?" 1996, http://www.wipp.carlsbad.nm.us/fctshts/factsheet.htm.

19. There have been a number of litigations against the DOE (by the State of New Mexico, the Environmental

Defense Fund, the Natural Resources Defense Council, the Southwest Research and Information Center, the Concerned Citizens for Nuclear Safety, and the State of Texas). And there have been actions against the EPA (by the State of New Mexico, the Southwest Research and Information Center, the Concerned Citizens for Nuclear Safety, and the Citizens for Alternatives to Radioactive Dumping). And against New Mexico, and so on.

20. See, for example, the Atomic Energy of Canada Limited Impact Statement for the now defunct Canadian disposal scenario. Atomic Energy Canada Limited and Ontario Hydro, *Environmental Impact Statement on the Concept for Disposal of Canada's Nuclear Fuel Waste* (Ottawa: Atomic Energy Canada, 1994).

21. An interesting twist to the question of the sterility of the salt environment appeared in the form of a letter to *Nature* magazine in fall 2000. The research reports having isolated several 250 million-year-old bacteria from salt crystals collected from one of the WIPP's air intake shafts. Brine inclusions within salt crystals are shown to contain a previously unknown spore-forming bacterium—the oldest known organism and arguably an endangered species! See Russell H. Vreeland, William D. Rosenzweig, and Dennis W. Powers, "Isolation of a 250 Million-Year-Old Halotolerant Bacterium from a Primary Salt Crystal," *Nature* 407 (October 19, 2000): 897–900. See also J. Travis, "Prehistoric Bacteria Revived from Buried Salt," *Science News* 155, no. 24 (1999): 373. This was not the first time such claims have been made, but it is being regarded as the most credible. See W. D. Grant, R. T. Gemmell, and T. J. McGenity, "Halobacteria: The Evidence for Longevity," *Extremophiles* 2 (1998): 279–87.

22. See U.S. Department of Energy, "Fact Sheet—Why Salt?" 1995, http://www.wipp.carlsbad.nm.us/fctshts/factsheet.htm.

23. See Human Interference Task Force, *Reducing the Likelihood of Future Human Activities That Could Affect Geologic High-Level Waste Repositories* (Columbus, OH: Office of Nuclear Waste Isolation, Battelle Memorial Institute, 1984).

24. Thomas A. Sebeok, *Communication Measures to Bridge Ten Millennia* (Columbus, OH: Office of Nuclear Waste Isolation, Battelle Memorial Institute, 1984), 24.

25. Ibid., 25.

26. Ibid.

27. Ibid., 27. Emphasis added.

28. A. Weitzberg, *Building on Existing Institutions to Perpetuate Knowledge of Waste Repositories* (Columbus, OH: Office of Nuclear Waste Isolation, Battelle Memorial Institute, 1982).

29. For a discussion of a Popperian approach to this problem, see Jan Nolin, "Communicating with the Future: Implications for Nuclear Waste Disposal," *Futures (London, England)* 25 (summer 1993): 778–91.

30. Percy H. Tannenbaum, *Communication across Three Hundred Generations: Deterring Human Interference with Waste Disposal Sites* (Columbus, OH: Office of Nuclear Waste Isolation, Battelle Memorial Institute, 1984).

31. Cf. Irenäus Eibl-Eibesfeldt, *Human Ethology* (New York: Aldine de Gruyter, 1989).

32. See Maureen Kaplan, *Archaeological Data as a Basis for Repository Marker Design*, ed. Analytical Sciences Corporation (Columbus, OH: Office of Nuclear Waste Isolation, Battelle Memorial Institute, 1982); and Maureen F. Kaplan and Mel Adams, "Using the Past to Protect the Future: Marking Nuclear Waste Disposal Sites," *Archaeology* 39 (Sept./Oct. 1986): 51–54. Aspects of Kaplan's work built on previous work of Givens. See D. B. Givens, "From Here to Eternity: Communicating with the Distant Future," *Et Cetera: A Review of General Semantics* 39, no. 2 (1982): 159–79.

33. Stephen J. Hora, Detlof von Winterfeldt, and Kathleen M. Trauth, *Expert Judgment on Inadvertent Human Intrusion into the Waste Isolation Pilot Plant* (Albuquerque: Sandia National Laboratories for the United States Department of Energy, 1991), I-2.

34. The group that was recruited was divided into four separate working groups or teams. Their reports are published separately as appendices to the Hora et al. report.

35. Ibid., I-2–I-3.

36. Ibid., C-40.

37. Thomas Kuhn, *The Structure of Scientific Revolutions* (Chicago: University of Chicago Press, 1970); Herbert Marcuse, *One-Dimensional Man* (Boston: Beacon Press, 1966).

38. For information on the Gnome detonation and Project Plowshares, see the Atomic Tourist site (http://www.atomictourist.com) and the Center for Land Use Interpretation (http://www.clui.org).

39. This site is worth a visit. Take Highway 285 south from Carlsbad, and go east on Route 31. Turn right on Route 128, and proceed east for 7.2 miles. Turn right on Mobley Ranch Road (also called Eddy County Road 795). After 0.6 miles, Mobley Ranch Road makes a sharp right—do not follow it. Instead, go straight through on the gravel road. Follow the gravel road for 4.3 miles, and then turn left onto the small dirt road. The monument is 150 feet away.

40. Hora et al., *Expert Judgment on Inadvertent Human Intrusion*, C-68.

41. The group recruited was divided into two teams. Team A: Dieter G. Ast (Cornell University), Michael

Brill (Buffalo Organization for Social and Technological Innovation), Ward Goodenough (University of Pennsylvania), Maureen Kaplan (Eastern Research Group), Frederick Newmeyer (University of Washington), and Woodruff Sullivan (University of Washington). Team B: Victor R. Baker (University of Arizona), Frank R. Drake (University of California at Santa Cruz), Ben R. Finney (University of Hawaii at Manoa), David B. Givens (American Anthropological Association), Jon Lomberg (independent artist, designer, and writer), Louis Narens (University of California at Irvine), and Wendell Williams (Case Western Reserve University).

42. Kathleen M. Trauth, Stephen C. Hora, and Robert V. Guzowski, *Expert Judgment on Markers to Deter Inadvertent Human Intrusion into the Waste Isolation Pilot Plant* (New Mexico: Sandia National Laboratories, 1993), F-27.

43. Once again, a contents bigger than its form. See §1.10.

44. Ibid., F-33 (emphasis in original).

45. Eibl-Eibesfeldt, *Human Ethology*. Eibl-Eibesfeldt, a student of Lorenz, is characteristically behaviorist in his approach to questions phylogenetic. The ethogram of man is a map upon which phylogenetic adaptation shapes and constrains behavior. In part a reaction to the extreme environmentalism—of which cultural relativism is held to be the zenith (or nadir)—Eibl-Eibesfeldt's ethographic work proceeds from the assumption that "man has been given some universal guide lines of how to behave," and that "[h]is inborn nature is the benchmark." Irenäus Eibl-Eibesfeldt, *Ethology: The Biology of Human Behavior*, trans. Erich Klinghammer (New York: Holt, Rinehart and Winston, 1975), 534.

46. See Eibl-Eibesfeldt, *Human Ethology*, and Eibl-Eibesfeldt, *Ethology*.

47. Gilles Deleuze and Félix Guattari, *A Thousand Plateaus*, trans. Brian Massumi (Minneapolis: University of Minnesota Press, 1987), 112.

48. Trauth et al., *Expert Judgment on Markers*, F-39.

49. Trauth et al., *Expert Judgment on Markers*, F-120. The (too clever) idea of representing the passage of time as a function of the precession of stars about Polaris appears to have originated with Carl Sagan.

50. Charles Saunders Peirce, *Collected Papers of Charles Sanders Peirce*, Volumes 1–8, ed. Charles Hartshorne and Paul Weiss (vols. 1–6) and Arthur W. Burks (vols. 7–8) (Cambridge, MA: Belknap Press of Harvard University Press, 1931–58), 2.92.

51. A more precise formulation here would be that the kind of sign and representation sought after with Mr. Yuk is on the order of a hypoicon. A "hypoicon may be roughly divided according to the mode of Firstness of which they partake. Those which partake of simple qualities, or First Firstnesses, are images; those which represent the relations, mainly dyadic, or so regarded, of the parts of one thing by analogous relations in their own parts, are diagrams; those which represent the representative character of a representamen by representing a parallelism in something else, are metaphors." Peirce, 2.277. What we have in the case of Mr. Yuk is a mostly diagrammatic hypoicon, where the figuration of the sign—its own distinctive points—are intended to map in a dyadic fashion to the characteristic points in an idea or possibility of nausea or sickness, and so on.

52. Peirce, 3.434.

53. Peirce, 1.92. He continues: "Any utterance of speech is an example. If the sounds were originally in part iconic, in part indexical, those characters have long since lost their importance. The words only stand for the objects they do, and signify the qualities they do, because they will determine, in the mind of the auditor, corresponding signs."

54. Peirce, 4.447.

55. Peirce, 3.291.

56. Ibid.

57. Carl Sagan, letter to Richard Anderson, Sandia National Laboratories, August 8, 1990. Reprinted in Trauth et al., *Expert Judgment on Markers*, G-88–G-89.

58. It is important to reflect somewhat on the gravity of introducing the thematic apperception test as an argument against the use of iconographic materials in the marker. As I understand it, the utility of the test is its ability to read the overdetermined expressive content of apperceptive distortion(s). It is not *simply* an example of the subjective manner in which signs are perceived. It certainly has something to say with respect to cultural aspects of perception, but as I understand it, it is not principally interested in the nomothetic content of apperception. Rather, its strength is as an idiographic instrument. See Leopold Bellak, "Theoretical Foundations for Projective Testing," in *The Thematic Apperception Test, The Child's Apperception Test, and the Senior Apperception Test in Clinical Use*, 3d ed. (New York: Grune and Stratton, 1975).

59. Trauth et al., *Expert Judgment on Markers*, F-44.

60. The English instruction, "This place is dangerous. Go away!" was entered into the Babelfish translator and then translated into another language, then back into English, then into another language, back into English, and so on. This is based on the procedure used by Marcos Novak. See Marcos Novak, "Next Babylon: Accidents to Play In," in *The Art of the Accident*, ed. Andreas Broeckmann, Joke Brouwer, Bart Lootsma, Arjen Mulder, and Lars Spuybroek (Rotterdam: NAi Publishers/V2_Organisatie, 1998).

61. Trauth et al., *Expert Judgment on Markers*, G-46.

62. See, for example, the proposal of Robert and Miriam Weisang Misrach for the creation of a 64-square-mile park of despoiled landscape. The area in question is the bombing range known as Bravo 20, a landscape bombed into an otherworldy toxicity by the U.S. Navy. The Misrach's national park would have interpretive programs, film and video archives, a "mess hall serving Navy chow," and a gift shop. Misrach and Misrach, *Bravo 20: The Bombing of the American West*. For commentary on the proposal see Miwon Kwon, "The Pleasures of Nature in Ruins," *Documents* 1, no. 1/2 (fall/winter 1992): 20–26; and Andrew Ross, "How to Occupy Your Own Country," *Documents* 1, no. 1/2 (fall/winter 1992): 12–19.

63. Trauth et al., *Expert Judgment on Markers*, F-52.

64. Ibid., F-57.

65. Ibid.

66. Ibid., F-58.

67. Ibid., F-136. Adding, with an inexplicable air of neighborly responsibility: "Noise levels would need to be controlled so as not to disturb people residing in the general vicinity."

68. The Futures Panel also identified sound as an important design element. See Hora et al., *Expert Judgment on Inadvertent Human Intrusion*, C-69.

69. The Southwest Research Information Center has been instrumental in organizing opposition and in undertaking oversight responsibilities. Information can be had from Southwest Research Information Center, PO Box 4524, Albuquerque, NM 87106. Phone: 505/262–1862. http://www.sric.org.

70. As of April 2003, the WIPP has received 1618 shipments of waste.

71. The name used for this sort of site is negotiated, monitored, retrievable storage (NMRS), and it has been unsuccessfully argued by critics of permanent geological storage. The most detailed treatment and critique of permanent geological storage that I have seen is K. S. Schrader-Frechette, *Burying Uncertainty: Risk and the Case against Geological Disposal of Nuclear Waste* (Berkeley: University of California Press, 1993). In a recent "Collective Opinion" rendered by the OECD Nuclear Energy Agency, they affirmed that disposal was consistent with global ethical principles for the future and, furthermore, that it is consistent with the vision of the future as set out in the Brundtland Commission report (World Commission on Environment and Development, *Our Common Future* [New York: Oxford University Press, 1987]). They stated that "[we] confirm that the geological disposal strategy can be designed and implemented in a manner that is sensitive and responsive to fundamental ethical and environmental considerations; [we] conclude that it is justified, both environmentally and ethically, to continue development of geological repositories for those long-lived radioactive wastes which should be isolated from the biosphere for more than a few hundred years; and [we] conclude that stepwise implementation of plans for geological disposal leaves open the possibility of adaptation, in the light of scientific progress and social acceptability, over several decades, and does not exclude the possibility that other options could be developed at a later stage." Radioactive Waste Management Committee OECD Nuclear Energy Agency, "The Environmental and Ethical Basis of the Geological Disposal of Long-Lived Radioactive Waste" (1994), http://193.51.64.1/html/rwm/reports/1995/geodisp.html.

72. U.S. Department of Energy, "Waste Isolation Pilot Plant Compliance Certification Application to the EPA" (1996), http://www.wipp.carlsbad.nm.us/library/cca/cca.htm

73. Ibid., section 13.

74. In keeping with earlier work on the marker, the DOE has elected to keep the number of monuments as a function of the power of 2.

75. "The conceptual configuration represents a simple design for ease of fabrication. Quarries contacted regarding the feasibility of producing large monoliths stated that fabrications of stone with flat surfaces were significantly simpler to quarry than those with curved surfaces. The wastage of material (and thus cost) is also significantly less for flat surface stones. Thus the choice was made to produce a foundation monolith in the shape of a truncated pyramid and a surface monolith as a regular four sided prism." http://reserve.wipp.carlsbad.nm.us/cgi-in/foliocgi.exe/App_pic.nfo/.

76. Ibid.

77. The shipping has not thus far been without incident. In November 2000, a waste truck from Idaho made a wrong turn heading for Albuquerque and ended up on a nonlicensed route. This should have been impossible with the real-time satellite tracking system that is administered in Oak Ridge, Tennessee. Nonetheless, the truck was eventually noticed and stopped by the local police department. In another incident, a truck was stopped and ticketed for not displaying proper information about its contents.

78. Environmental Protection Agency, "Passive Institutional Controls, CARD 43" (1998), http://www.epa.gov/radiation/wipp/docs/card43.pdf.

Threat and Trauma

1. Toronto filmmaker and director Peter Blow's remarkable documentary, *Village of Widows*, tracks an instance of escape and confession. This extremely powerful short film follows an ethical line of implication or flight from the Déné of Great Bear Lake, through Port Hope, Ontario, the Manhattan Project, to Hiroshima. The Déné continue to endure the loss of many to radiation-related sickness—attributable to their employment (between 1932 and 1960) transporting radioactive ore (radium and uranium) from a mine site at Port Radium, Northwest Territories. This material was moved over a 2,500 km water route to the railhead in northern Alberta, where it was transported south to the refinery in Port Hope, Ontario, and then taken to the United States where it (together with ore from the Belgian Congo and Colorado) was made into the bombs (the Manhattan Project) that were detonated over Hiroshima and Nagasaki. In the wake of their history of traumatic loss, and within their ongoing suffering, the Déné reached the remarkable conclusion that they too were complicit in the bombing of Japan. Their ethical obligation was to make an extraordinary visit to Hiroshima to tell the Japanese survivors that they were sorry, that they didn't know, and that they too would commemorate the stunning loss of life. This history is set out as a problem of and for memory and landscape in my "The Highway of the Atom: Recollections along a Route," *Topia* 7 (2002).

2. Harold Osborne, ed., *The Oxford Companion to Art* (London: Oxford University Press, 1970), 737.

3. Here I think of the Gerzes's disappearing "Harburg Monument Against Fascism," detailed in James E. Young, "The Counter-Monument: Memory against Itself in Germany Today," *Critical Inquiry* 18, no. 2 (1992): 267–96.

4. "In Germany, the Holocaust signifies an absence of Jews and a traumatic burden on the national identity, in which genuine attempts at mourning are hopelessly entangled with narcissistic injury, ritual breast-beating, and repression.... In Israel, the Holocaust became central to the foundation of the state, both as the end point of a disavowed history of Jews as victims, *and* as a starting point for a new history of nation, self-assertion, and resistance.... The American focus of the Holocaust concentrates on America as liberator of the camps and haven for refugees and immigrants." Andreas Huyssen, "Monument and Memory in a Postmodern Age," in *The Art of Memory: Holocaust Memorials in History*, ed. James E. Young (Munich: Prestel-Verlag, 1994), 13–14.

5. Cf. Maureen Kaplan, *Archaeological Data as a Basis for Repository Marker Design*, ed. Analytical Sciences Corporation (Columbus, OH: Office of Nuclear Waste Isolation, Battelle Memorial Institute, 1982); and Human Interference Task Force, *Reducing the Likelihood of Future Human Activities That Could Affect Geologic High-Level Waste Repositories* (Columbus, OH: Office of Nuclear Waste Isolation, Battelle Memorial Institute, 1984).

6. Writing in Germany in the wake of the Chernobyl disaster, Ulrich Beck relates this idea of perceptual

expropriation to a larger sense of "anthropological shock." Ulrich Beck, *Ecological Enlightenment: Essays on the Politics of the Risk Society*, trans. Mark A. Ritter (Atlantic Highlands, NJ: Humanities Press International, 1995), 63–76.

7. François Ewald, "Insurance and Risk," in *The Foucault Effect: Studies in Governmentality*, ed. Graham Burchell, Colin Gordin, and Peter Miller (Chicago: University of Chicago Press, 1991), 199. Emphasis added.

8. There are a host of competing statistics that one could invoke here. However, since the most common statistics are keyed to insured losses, they tend to be economically skewed. The numbers here, while noncanonical, represent economic loss, irrespective of the value of insured loss. The German reinsurance company, Munich Re, regularly compiles and publishes global synopses of economic disaster loss. See, for example, Geoscience Research Group of Munich Re, *World Map of Natural Hazards* (Munich: Münchener Rüchversicherungs-Gesellschaft, 1998).

9. See Mike Davis, *Ecology of Fear: Los Angeles and the Imagination of Disaster* (New York: Metropolitan Books, 1998), 32–33.

10. Kenneth A. Froot, introduction to *The Financing of Catastrophe Risk* (Chicago: University of Chicago Press, 1999).

11. François Ewald, "Two Infinities of Risk," in *The Politics of Everyday Fear*, ed. and trans. Brian Massumi (Minneapolis: University of Minnesota Press, 1993), 222.

12. Ibid.

13. That such threats appear as artifice strikes me as a measure of an insistence that threats are somehow *unexpected*. I prefer Virilio's term, *inverted miracle*. See Paul Virilio, *Politics of the Very Worst: An Interview by Phillipe Petit*, ed. Sylvère Lotringer, trans. Michael Cavaliere (New York: Semiotext[e], 1999), 89.

14. For example, Ulrich Beck levels this charge against Niklas Luhmann. He claims that Luhmann's assent to living with the virtuality of catastrophe amounts to delivering sociological enlightenment to the hands of an industrial and technological teleology. He asks if sociological enlightenment will not—in the face of potential catastrophe—become, *mutatis mutandis*, an industrial fatalism. Beck, *Ecological Enlightenment*, 113–14.

15. Ewald, "Two Infinities of Risk," 224.

16. Ibid.

17. Ibid., 225.

18. Ibid.

19. To say nothing of the near "natural" connections between survivalist movements and concepts of purity (racial, social), propped up by various Christian and apocalyptic rhetorics. To be sure, the European imaginary contains some odd conceptions of the American social landscape (pace Baudrillard via de Tocqueville). But to take him at his word, Ewald's totalizing of threat can perhaps too easily end up validating any consensus, any capture of a new identity, "a new skin," as a revolutionary practice.

20. Quoted in Lionel Rubinoff, *The Pornography of Power* (Chicago: Quadrangle Books, 1968). See Kenneth Rexroth, "Disengagement: The Art of the Beat Generation," in *The Beat Generation and the Angry Young Men*, ed. Gene Feldman and Max Gartenberg (New York: Dell, 1959).

21. See Bikenibeu Paeniu, "Tuvalu Statement Presented by His Excellency the Rt Hon Bikenibeu Paeniu, Prime Ministerial Special Envoy on Climate Change, United Nations Framework Convention on Climate Change—Conference on Parties 3," December 8, 1997, http://www.tuvaluislands.com/kyoto-panieu.htm.

22. Here, the metaphor of the shipwreck and the spectator becomes a great deal less metaphorical. See Hans Blumenberg, *Shipwreck with Spectator: Paradigm of a Metaphor for Existence*, trans. Steven Rendall (Cambridge: MIT Press, 1997).

23. See Jason Burke, "After the Cyclone Comes the Anger," *Observer/Guardian Weekly*, November 11, 1999.

24. See Mike Davis, *Ecology of Fear: Los Angeles and the Imagination of Disaster* (New York: Metropolitan Books, 1998), 151–94. Davis's writing here—as with his earlier *City of Quartz: Excavating the Future in Los Angeles* (New York: Vintage Books, 1992)—can be read as primary fieldwork in the sociology and philosophy of threat. Davis's California is *the* contemporary theme park of disavowal.

25. See http://www.tv. Tuvalu is apparently no stranger to renting and selling off bits of its technological apparatus. They have made a considerable sum from leasing their international area code to phone sex operations. Currently they are negotiating with Australia for a home for what is perhaps the world's first group of global warming refugees.

26. See Faisal Islam, "Net Profit Sees Tuvalu into UN," *The Guardian Weekly*, September 19, 2000, 5.

27. Beck, *Ecological Enlightenment*, 122. See also Ulrich Beck, *Ecological Politics in an Age of Risk*, trans. Amos Weisz (Cambridge, U.K.: Polity Press, 1995), esp. 158–84.

28. Beck, *Ecological Enlightenment*, 2.

29. Ibid.

30. Claus Offe, quoted in Beck, *Ecological Enlightenment*, 3.

31. Beck, *Ecological Enlightenment*, 3.

32. Beck here is concerned with the manner in which threats ultimately impinge upon rights. Indeed, a good deal of his critique is contained within a discourse of rights vis-à-vis the freedoms upon which threat impinges. Perhaps this is necessary in the sense that it gives a kind of ground upon which to contest the global and other practices that are actively engaged in the manufacture of these threats, but it falls short of dealing on a more fundamental level with what to do in the face of these threats that are, so to speak, performative.

33. Slavoj Žižek, *Looking Awry: An Introduction to Jacques Lacan through Popular Culture* (Cambridge: MIT Press, 1991), 35.

34. At the limit, the logic of capitalism itself may be seen as a massive disavowal of the material constraints of organic life. Which is to say the "real" dependent hierarchy, from land to labor to capital, is disavowed in an imaginary (inverted) hierarchy of capital to labor to land. See Anthony Wilden's brilliant analysis in *System and Structure: Essays in Communication and Exchange*, 2d ed. (New York: Tavistock, 1980), xxxi–xxxvii.

35. For Freud there was a kinship between disavowal and psychosis that does not exist (in the same fashion) with neurosis. The very movement of disavowal is what founds psychosis, and what effectively creates an opposition between psychosis and neurosis. In both cases, as Freud described it (1924) there is a loss of reality, a rebellion of the id against the world. And in both cases it is in relation to the id that a negotiation takes place—a preponderance of reality over id in the case of neurosis, and the reverse in psychosis. "In neurosis a part of reality is avoided by a sort of flight, but in psychosis it [reality] is remodeled." Sigmund Freud, "The Loss of Reality in Neurosis and Psychosis" (1924), in *Collected Papers*, Volume 5, trans. Joan Riviere (New York: Basic Books, 1959), 279. Which is to say that in the neurotic response to the ecological threat there is no attempt to disavow a reality, but merely to ignore it through a repression. While on the other hand, the psychotic response is to disavow it entirely *and* put something else in its place, to refashion a reality through the active construction of perception and falsification of memory (that is, "by creating a new reality which is no longer open to objections like that which has been forsaken"). Ibid.

36. Julia Kristeva, *New Maladies of the Soul*, trans. Ross Guberman (New York: Columbia University Press, 1995), 46–49.

37. Ibid., 49. On the displacement strategies of the obsessional neurotic see Sigmund Freud, "Notes upon a Case of Obsessional Neurosis" (1909), in *Collected Papers*, Volume 3, trans. Alix and James Strachey (New York: Basic Books, 1959), 334–36.

38. This theme is considered at some length in my *Primitives in the Wilderness* (Albany: State University of New York Press, 1997).

39. Žižek, *Looking Awry*, 36.

40. Jacques Lacan, *The Seminar of Jacques Lacan: Book I: Freud's Papers on Technique, 1953–1954*, trans. John Forrester (New York: Norton, 1988), 97.

41. Ibid., 66.

42. See, for example, Jacques Lacan, "Seminar on 'The Purloined Letter,'" in *The Purloined Poe: Lacan, Derrida, and Psychoanalytic Reading*, ed. John P. Muller and William J. Richardson, trans. Jeffrey Mehlman (Baltimore: Johns Hopkins University Press, 1988). See also Bruce Fink, "The Nature of Unconscious Thought or Why No One Ever Reads Lacan's Postface to the 'Seminar on "The Purloined Letter,"'" in *Reading Seminars I and II: Lacan's Return to Freud*, ed. Richard Feldstein, Bruce Fink, and Maire Jaanus (Albany: State University of New York Press, 1996). Fink shows what Lacan may have meant by his contention that the raw events of the world are never innocent, never—well, almost never—untouched by the symbolic. "We assume the real event in question—the tossing of the coin—to be *random*, that is, we presuppose that the coin was not loaded. But what does it mean for a coin *not* to be loaded? Generally it means that it is exactly as likely to turn up heads as it is tails. How is this determined? By throwing it over and over, and counting the number of times each possibility turns up, an acceptable coin being one which, out of 1000 tosses, gives us 500 heads and 500 tails. Which is tantamount to saying that it is our already existing symbolic system which determines whether the event in question is considered *random* or not.... Which is to say that the 'raw event' with which we began was already symbolically determined, and that the symbolic matrices are never 'innocent,' that is never lacking in incidence on our supposedly 'pregiven reality.' The event is thus retroactively constituted as random by the signifier." Fink, "The Nature of Unconscious Thought," 188–89.

43. Jacques Lacan, "Odd or Even? Beyond Intersubjectivity," in *The Seminar of Jacques Lacan: Book II: The Ego in Freud's Theory and in the Techniques of Psychoanalysis, 1954–1955*, trans. Sylvana Tomaselli (New York: Norton, 1988), 182.

44. And it may well irrupt into reality again. See "The Chernobyl Report: The Site and Accident Sequence" (OECD Nuclear Energy Agency, 1996), vol. 2000.

45. Vladimir Shevchenko, *Chernobyl: Chronicle of Difficult Weeks*, VHS, The Video Project, Oakland, CA, 1986.

46. Slavoj Žižek, *The Sublime Object of Ideology* (London: Verso, 1989), 163.

47. Bon mot of Montréal psychoanalyst and teacher Charles Levin.

48. See Bruce Fink, *The Lacanian Subject: Between Language and Jouissance* (Princeton, NJ: Princeton University Press, 1995), 27.

49. Peirce, 6.344.

50. Jacques Lacan, *The Seminar of Jacques Lacan: Book VII: The Ethics of Psychoanalysis, 1959–1960*, trans. Dennis Porter (New York: Norton, 1992), 240.

51. In reference to the Revelation of St. John the Divine, one might actually count to three: a spiritual death (a separation from god), a physical death (of the body), and a second death (which annihilates the very cycle of death and resurrection).

52. As Žižek has it, the (our) relation to Chernobyl is given as the formula for phantasy: $\$ \lozenge a$. And I suppose this is the case providing that what we see in Chernobyl is a *nature disrupted*. Lacan attributes the formula to Jean-Claude Milner (see Jacques Lacan, *Écrits: A Selection*, trans. Alan Sheridan [New York: Norton, 1977], 334), and Wilden explains it thus: The $\lozenge$ refers to the relationships of envelopment-development-conjunction-disjunction, and the $\$$ refers to the Other subject in the subject's division from himself. The a denotes an object of identification. See Jacques Lacan and Anthony Wilden, *Speech and Language in Psychoanalysis*, trans. Anthony Wilden (Baltimore: Johns Hopkins University Press, 1968), 267.

53. Žižek, *Looking Awry*, 37.

54. Ibid., 38.

55. The displacement of threat into a prior condition of disconnection deflects threat in the same manner as, for example, Blanchot's disaster is displaced in relation to threat. Blanchot writes: "The disaster, depriving us of that refuge which is the thought of death, dissuading us from the catastrophic or the tragic, dissolving our interest in will and all eternal movement, does not allow us to entertain this question either: what have you done to gain knowledge of the disaster?" Maurice Blanchot, *The Writing of the Disaster*, trans. Ann Smock (Lincoln: University of Nebraska Press, 1986), 3.

56. Lacan, *The Ethics of Psychoanalysis*, 104.

57. Ibid., 36.

58. Ibid., 20.

59. Ibid., 21.

60. As I wrote an early draft of this section in a motel room in Bowling Green, Ohio, there was a National Public Radio test of the emergency broadcast system: "Had this been a real emergency," it interrupted in order to warn me, "a tornado, a toxic spill, or a nuclear accident, it—," the interruption thus interrupted, it cut back to a program about the growth of HIV/AIDS in sub-Sahara Africa.

61. Sigmund Freud, "Beyond the Pleasure Principle" (1920), in *On Metapsychology: The Theory of Psychoanalysis,* Volume 11, Pelican Freud Library, ed. and trans. James Strachey (New York: Penguin Books, 1984), 281–82.

62. See Ruth Leys, *Trauma: A Genealogy* (Chicago: University of Chicago Press, 2000).

63. Freud, "Beyond the Pleasure Principle," 301.

64. Freud had previously concerned himself with trauma in his work with Breuer; see Josef Breuer and Sigmund Freud, *Studies on Hysteria (1893–1895),* ed. and trans. James Strachey (New York: Basic Books, 1957). See also his 1894 paper, "The Defense of Neuropsychosis," in Sigmund Freud, *Collected Papers,* Volumes 1–5, trans. Joan Riviere (New York: Basic Books, 1959).

65. Slavoj Žižek, "Hegel with Lacan, or the Subject and Its Cause," in *Reading Seminars I and II: Lacan's Return to Freud,* ed. Richard Feldstein, Bruce Fink, and Maire Jaanus (Albany: State University of New York Press, 1996), 399.

66. Ibid., 398. So again, it is through the symbolic that the real appears. And though (in his first seminar) Lacan has said that the real itself, however, is without gaps, without fissure, he will change this position to say that there are indeed holes in the real; holes that result from the fashioning of the signifier, and from such events as trauma. On this point about the real see *The Seminar of Jacques Lacan: Book II: The Ego in Freud's Theory and in the Techniques of Psychoanalysis,* 97–98. There is a kind of creation, *ex nihilo,* in which something is fashioned from the emptiness. And the production of a hole creates, simultaneously, the possibility for filling it. The analogy Lacan uses is that of the potter and the vase: "It [the vase] creates the void and thereby introduces the possibility of filling it. Emptiness and fullness are introduced into a world that by itself knows not of them. It is on the basis of this fabricated signifier, this vase, that emptiness and fullness as such enter the world." *The Ethics of Psychoanalysis,* 120. In the case of the traumatic event, the hole in the real, and the corresponding gap or disturbance in the symbolic, take on the status for Lacan of a Thing, or *das Ding.* "*Das Ding* is at the center only in the sense that it is excluded. That is to say, in reality *das Ding* has to be posited as exterior, as the prehistoric Other that it is impossible to forget—the Other whose primacy of position Freud affirms in the form of something *entfremdet,* something strange to me, although it is at the heart of me, something that on the level of the unconscious only a representation can represent." *The Ethics of Psychoanalysis,* 71. *Das Ding,* then, as the place that cannot be symbolized, yet is produced by the very operation of symbolization ("it secretes, it isolates the empty 'in-

digestible' place of the Thing"), brings us to the point where we can see that trauma comes very close to the idea of the real itself; this missed event that fails to be symbolically integrated.

67. Sigmund Freud, "Analysis of a Phobia in a Five-Year-Old Boy" (1909), in *Collected Papers,* Volume 3, trans. Alix and James Strachey (New York: Basic Books, 1959), 263–64. Quoted in Jean Laplanche and J.-B. Pontalis, *The Language of Psychoanalysis,* trans. Donald Nicholson-Smith (New York: Norton, 1973), 79.

68. Sigmund Freud, "Project for a Scientific Psychology" (1895), in *Origins of Psychoanalysis: Sigmund Freud's Letters to Wilhelm Fliess, 1887–1902,* ed. Marie Bonaparte, Anna Freud, and Ernst Kris, trans. Eric Mosbacher and James Strachey (New York: Basic Books, 1954), 413, quoted in Jean Laplanche, *Life and Death in Psychoanalysis,* trans. Jeffrey Mehlman (Baltimore: Johns Hopkins University Press, 1976), 41.

69. Laplanche, *Life and Death in Psychoanalysis,* 41.

70. On the question of trauma's mode of derivation, see Laplanche, "Appendix: The Derivation of Psychoanalytic Entities," in ibid. In one sense, it is to situate the psychical trauma as a prolongation or continuation of the medicosurgical theory of the physical trauma. In another sense, it is to transpose more or less the elements of the later—the physical trauma—into a different sphere altogether. "That of an extension through *continuity,* an imperceptible transition to an adjacent field; and that of a transposition through *similarity* into a field that is different but structured as analogous" (ibid., 131). In the trauma literature it is not always clear which of these two modes is prominent.

71. For example, see the classic encyclopedia entry: Charles E. Fritz, "Disaster," in *International Encyclopedia of the Social Sciences,* ed. David L. Sill, vol. 4 (New York: Macmillan, 1963).

72. See Claude Gilbert, "Studying Disaster: Changes in the Main Conceptual Tools," in *What Is a Disaster? Perspectives on the Question,* ed. E. L. Quarantelli (New York: Routledge, 1998), 11–18. As far as I know, he first proposed the categorization of disaster discussed in the chapter text.

73. See Michel Serres, *The Natural Contract,* trans. Elizabeth MacArthur and William Paulson (Ann Arbor: University of Michigan Press, 1995).

74. As the co-director of the Katastrophenforschungsstelle at Christian Albrechts University in Germany, Wolf Dombrowsky, put it.

75. Kenneth Hewitt, "Excluded Perspectives in the Social Construction of Disaster," in Quarantelli, *What Is a Disaster?,* 80.

76. Mary Wolfenstein, *Disaster: A Psychological Essay* (London: Routledge & Kegan Paul, 1957), 189.

77. Kai T. Erikson, *Everything in Its Path: Destruction of Community in the Buffalo Creek Flood* (New York: Simon and Schuster, 1976).

78. Ibid., 255.

79. Ibid.

80. Ibid., 258. Erikson continued work on areas of disaster and has published a collection of essays concerning the social and ecological disasters of Grassy Narrows, Immokalee, Three Mile Island, the Yucca Mountain project, Hiroshima, and the homeless in America. See Erikson, *A New Species of Trouble: The Human Experience of Modern Disasters* (New York: Norton, 1994).

81. Sigmund Freud, "Civilization and Its Discontents," in *Civilization, Society and Religion, Group Psychology, Civilization and Its Discontents, and Other Works*, Volume 12, Pelican Freud Library, trans. James Strachey (New York: Penguin Books, 1985), 338, emphasis added. See also my "The Highway of the Atom: Recollections Along a Route" *Topia* 7 (spring 2002): 99–115.

82. Robert D. Kaplan, *The Ends of the Earth: A Journey at the Dawn of the 21st Century* (New York: Random House, 1996). Kaplan, an American journalist and travel writer, wrote *Ends of the Earth* based on his travels through Western and Northern Africa, parts of the Middle East, Eurasia, and China. On one hand, this work can be read as a documentary of how eco-social disaster constitutes the *umwelt* and *oikos* of the regions in which he traveled. And on the other hand, one can read this as a documentary account of how one neoconservative American journalist's neo-Malthusian expectations are rendered absurd in the face of the complexity he confronts.

83. Lifton discusses this feature of inquiry into trauma in relation to his work on Nazi doctors in his interview with Caruth. See Lifton, "An Interview with Robert J. Lifton," in *Trauma: Explorations in Memory*, ed. Cathy Caruth (Baltimore: Johns Hopkins University Press, 1995), 128–50.

84. Dominick LaCapra, "Lanzmann's *Shoah*: 'Here There Is No Why,'" in *History and Memory after Auschwitz* (Ithaca: Cornell University Press, 1998). Lanzmann's film *Shoah* is certainly a film about life and about death and about memory. What makes this film work has little to do with history and much to do with memory. Indeed, Lanzmann himself says that "Shoah is not a documentary ... [T]he film is not at all representational." Lanzmann has also said that *Shoah* is "a fiction of the real." Claude Lanzmann, "The Obscenity of Understanding: An Evening with Claude Lanzmann," in Caruth, *Trauma*. "Positivism and objectivism, which Lanzmann clearly rejects, deny or repress a transferential relation to the object whereby crucial aspects of it are repeated in the discourse or experience of the observer.

In acting out, on the contrary, one reincarnates or relives the past in an unmediated transferential process that subjects one to possession by haunting objects and to compulsively repeated incursions of traumatic residues (hallucinations, flashbacks, nightmares). Here the quest for full presence becomes phantasmatic and entirely uncontrolled." LaCapra, "Lanzmann's *Shoah*."

85. Laplanche and Pontalis, *The Language of Psychoanalysis*, 488–89.

86. This term, traumatic transvaluation, was suggested by Dominick LaCapra in a conversation with me in 1996. I am not sure if he has used it elsewhere.

87. Cf. Bruce Fink, *A Clinical Introduction to Lacanian Psychoanalysis: Theory and Technique* (Cambridge: Harvard University Press, 1997), 3–27.

88. Gilles Deleuze, "How Do We Recognize Structuralism," in *The Two-Fold Thought of Deleuze and Guattari: Intersections and Animations*, ed. Charles Stivale, trans. Melissa McMahon and Charles Stivale (New York: Guilford Press, 1998), 267.

89. Gilles Deleuze, *Difference and Repetition*, trans. Paul Patton (New York: Columbia University Press, 1994), 208–9.

90. "The possible has no reality (although it may have an actuality); conversely the virtual is not actual, *but as such possesses a reality* ... on the other hand ... the possible is that which is realized (or is not realized)." Gilles Deleuze, *Bergsonism*, trans. Hugh Tomlinson and Barbara Habberjam (New York: Zone Books, 1988), 96–97.

91. This is but a sleight of hand, wrote Deleuze. "In fact it is not the real that resembles the possible, it is the possible that resembles the real, because it has been abstracted from the real once made, arbitrarily extracted from the real like a sterile double." Ibid., 98.

92. Deleuze, *Difference and Repetition*, 212.

93. Deleuze, *Bergsonism*, 97.

94. This is all set out very clearly in ibid., 15–18, 94–98. See also Hardt's careful commentary on this aspect of Deleuze's work, which is most helpful. See Michael Hardt, *Gilles Deleuze: An Apprenticeship in Philosophy* (Minneapolis: University of Minnesota Press, 1993).

95. Deleuze, *Bergsonism*, 98.

96. Deleuze, *Difference and Repetition*, 211.

97. I am aware of Badiou's disputatious, duck-rabbit text on the question of Deleuze's aristocratic philosophy—Alain Badiou, *Deleuze: The Clamor of Being*, trans. Louise Burchill (Minneapolis: University of Minnesota Press, 2000)—I doubt he would agree with my reading here.

98. Deleuze, *Bergsonism*, 98.

99. See Brian Massumi, *A User's Guide to Capitalism and Schizophrenia: Deviations from Deleuze and Guattari* (Cambridge: MIT Press, 1992), 35–37.

100. See Deleuze, "How Do We Recognize Structuralism," 268.

101. Ibid., 270.

102. Paul Virilio, "The Primal Accident," in *The Politics of Everyday Fear*, ed. and trans. Brian Massumi (Minneapolis: University of Minnesota Press, 1993), 216.

103. Ibid. Emphasis added.

104. Hanna Segal puts this point much more categorically: "In natural death, or even in conventional war, men die, or at least those who have acquired some maturity die, with some conviction of symbolic survival in their children, grandchildren, in their work, or in the civilization of which they were part. Also, mourning, reconciliation and reparation can eventually take place.... The existence of nuclear weapons and the prospect of nuclear war make impossible the growing acceptance of death and symbolic survival." Hanna Segal, "From Hiroshima to the Gulf War and After: A Psychoanalytic Perspective," in *Psychoanalysis in Contexts: Paths between Theory and Modern Culture*, ed. Anthony Elliot and Stephen Frosh (New York: Routledge, 1995).

105. In "The Primal Accident," Virilio observes that at the end of the Second World War the critical boundary or threshold between the presence of weapons and their deployment is mirrored in the case of a nuclear power plant in the threshold between the function of chance and the crossing of that threshold to the accident.

106. Michel Foucault, *The Archaeology of Knowledge, and The Discourse on Language*, trans. A. M. Sheridan Smith (New York: Pantheon, 1972), 228.

107. See, for example, the Nuclear Guardianship Ethic. It is a set of pragmatic, ethical principles that seek to situate the extreme toxicity and longevity of nuclear materials into custodial practices of memory. See Nuclear Guardianship Project, "The Nuclear Guardianship Ethic," 1994, http://www.ratical.org/radiation/ NGP/Ngethic.html.

108. See Søren Kierkegaard, *The Concept of Anxiety: A Simple Psychologically Orienting Deliberation on the Dogmatic Issue of Hereditary Sin*, trans. Reidar Thomte (Princeton, NJ: Princeton University Press, 1980). "If I am anxious about a past misfortune, then this is not because it is in the past but because it may be repeated, i.e., become future" (91).

109. Jacques Derrida, "Whom to Give to (Knowing Not to Know)," in *The Gift of Death*, trans. David Wills (Chicago: University of Chicago Press, 1995), 54, 53. Derrida's pages devoted to Kierkegaard are some of the most beautiful he has written.

Index

Peter C. van Wyck is associate professor of communication at Concordia University in Montréal. He is the author of *Primitives in the Wilderness: Deep Ecology and the Missing Human Subject.*